A HANDBOOK OF PROOFS AND THEOREMS

Edited by
Maria Catherine C. Borres

ArclerPress

A Handbook of Proofs and Theorems

Edited by Maria Catherine C. Borres.

Edition 2018

ArclerPress

Published by
Arcler Press LLC
708 3rd Avenue, 6th Floor
New York
NY 10017
United States of America
www.arclerpress.com

ISBN: 978-1-68094-588-1 (Hardback)

Printed in United States of America

Library of Congress Control Number: 2017933179

About the Editor

Maria Catherine C. Borres

Catherine is currently taking up Master of Arts in Education Major in Mathematics in Philippine Normal University – Manila. She is currently working as a Content Developer for Mathematics at the Affordable Private Education Center (APEC Schools).

Table of Contents

2 MATHEMATICAL INDUCTION

3 MATHEMATICAL PROOF

4 AXIOMS

5 CONJECTURE

6 MISNAMED THEOREMS

7 PROPOSITION

8 INEQUALITY MATHEMATICS

9 FUNDAMENTAL THEOREMS

Preface

A theorem is a statement that can be demonstrated to be true by accepted mathematical operations and arguments. In general, a theorem is an embodiment of some general principle that makes it part of a larger theory. The process of showing a theorem to be correct is called a proof. Proofs employ logic but usually include some amount of natural language which usually admits some ambiguity. In fact, the vast majority of proofs in written mathematics can be considered as applications of rigorous informal logic. Purely formal proofs, written in symbolic language instead of natural language, are considered in proof theory. The distinction between formal and informal proofs has led to much examination of current and historical mathematical practice, quasi-empiricism in mathematics, and so-called folk mathematics. Many mathematical theorems are conditional statements. In this case, the proof deduces the conclusion from conditions called hypotheses or premises. In light of the interpretation of proof as justification of truth, the conclusion is often viewed as a necessary consequence of the hypotheses, namely, that the conclusion is true in case the hypotheses are true, without any further assumptions. However, the conditional could be interpreted differently in certain deductive systems, depending on the meanings assigned to the derivation rules and the conditional symbol. The philosophy of mathematics is concerned with the role of language and logic in proofs, and mathematics as a language.

Content Coverage

Chapter One focuses on history of logic that deals with the study of the development of the science of valid inference (logic).

Chapter Two presents an overview of a mathematical proof technique used to prove a given statement about any well-ordered set. Most commonly, it is used to establish statements for the set of all natural numbers.

Chapter Three focuses on proofs that employ logic but usually include some amount of natural language which usually admits some ambiguity.

Chapter Four focuses on axioms that serves as a starting point from which other statements are logically derived.

Chapter Five discusses about the conjecture, a conclusion or proposition based on incomplete information, for which no proof has been found. Conjectures such as the Riemann hypothesis or Fermat's Last Theorem have shaped much of mathematical history as new areas of mathematics are developed in order to prove them.

Chapter Six is focused on the misnamed theorems in mathematics. It includes theorems (and lemmas, corollaries, conjectures, laws, and perhaps even the odd object) that are well known in mathematics, but which are not named for the originator.

Chapter Seven identifies the propositions that are the sharable objects of attitudes and the primary bearers of truth and falsity. This stipulation rules out certain candidates for propositions, including thought- and utterance-tokens which are not sharable, and concrete events or facts, which cannot be false.

Chapter Eight focuses on inequalities used to demonstrate instances where one value is not the same as another value. The five symbols are described as "not equal to," "greater than," "greater than or equal to," "less than" and "less than or equal to.

Chapter Nine presents an emphasis on fundamental theorems of mathematics considered central to that field. The naming of such a theorem is not necessarily based on how often it is used or the difficulty of its proofs.

CHAPTER 1

History of Logic

Chapter Contents

- **Introduction**
- **Logic in the West**
- **Medieval Logic**
- **Traditional Logic**
- **Modern Logic**
- **Logic in the East**

The history of logic documents the development of logic as it occurs in various cultures and traditions in history. While many cultures have employed intricate systems of reasoning, logic as an explicit analysis of the methods of reasoning received sustained development originally only in three traditions: China, India and Greece. Although exact dates are uncertain, especially in the case of India, it is possible that logic emerged in all three societies in the fourth century B.C.E. The notions of systems of reasoning and logic, however, are sufficiently imprecise that various answers to the questions of what they are and how they are to be understood have been given. The formally sophisticated treatment of modern logic descends from the Greek tradition, but comes not wholly through Europe, but instead comes from the transmission of Aristotelian logic and commentary upon it by Islamic philosophers to logicians in Medieval Europe.

INTRODUCTION

The history of logic deals with the study of the development of the science of valid inference (logic). Formal logics developed in ancient times in China, India, and Greece.

Greek methods, particularly Aristotelian logic (or term logic) as found in the Organon, found wide application and acceptance in Western science and mathematics for millennia. The Stoics, especially Chrysippus, began the development of predicate logic.

Christian and Islamic philosophers such as Boethius (died 524) and William of Ockham (died 1347) further developed Aristotle's logic in the Middle Ages, reaching a high point in the mid-fourteenth century. The period between the fourteenth century and the beginning of the nineteenth century saw largely decline and neglect, and at least one historian of logic regards this time as barren. Empirical methods ruled the day, as evidenced by Sir Francis Bacon's Novum Organon of 1620.

Logic revived in the mid-nineteenth century, at the beginning of a revolutionary period when the subject developed into a rigorous and formal discipline which took as its exemplar the exact method of proof used in mathematics, a hearkening back to the Greek tradition. The development of the modern "symbolic" or "mathematical" logic during this period by the likes of Boole, Frege, Russell, and Peano is the most significant in the two-thousand-year history of logic, and is arguably one of the most important and remarkable events in human intellectual history.

Progress in mathematical logic in the first few decades of the twentieth century, particularly arising from the work of Gödel and Tarski, had a significant impact on analytic philosophy and philosophical logic, particularly from the 1950s onwards, in subjects such as modal logic, temporal logic, deontic logic, and relevance logic.

LOGIC IN THE WEST

Prehistory of Logic

Valid reasoning has been employed in all periods of human history. However, logic studies the *principles* of valid reasoning, inference and demonstration. It is probable that the idea of demonstrating a conclusion first arose in connection with geometry, which originally meant the same as "land measurement". The ancient Egyptians discovered geometry, including the formula for the volume of a truncated pyramid. Ancient Babylon was also skilled in mathematics. Esagil-kin-apli's medical *Diagnostic Handbook* in the 11th century BC was based on a logical set of axioms and assumptions, while Babylonian astronomers in the 8th and 7th centuries BC employed an internal logic within their predictive planetary systems, an important contribution to the philosophy of science.

Ancient Greece before Aristotle

While the ancient Egyptians empirically discovered some truths of geometry, the great achievement of the ancient Greeks was to replace empirical methods by demonstrative proof. Both Thales and Pythagoras of the Pre-Socratic philosophers seem aware of geometry's methods.

Fragments of early proofs are preserved in the works of Plato and Aristotle, and the idea of a deductive system was probably known in the Pythagorean school and the

Platonic Academy. The proofs of Euclid of Alexandria are a paradigm of Greek geometry. The three basic principles of geometry are as follows:

Certain propositions must be accepted as true without demonstration; such a proposition is known as an axiom of geometry.

- Every proposition that is not an axiom of geometry must be demonstrated as following from the axioms of geometry; such a demonstration is known as a proof or a "derivation" of the proposition.
- The proof must be *formal*; that is, the derivation of the proposition must be independent of the particular subject matter in question.

Further evidence that early Greek thinkers were concerned with the principles of reasoning is found in the fragment called *dissoi logoi*, probably written at the beginning of the fourth century BC. This is part of a protracted debate about truth and falsity. In the case of the classical Greek city-states, interest in argumentation was also stimulated by the activities of the Rhetoricians or Orators and the Sophists, who used arguments to defend or attack a thesis, both in legal and political contexts.

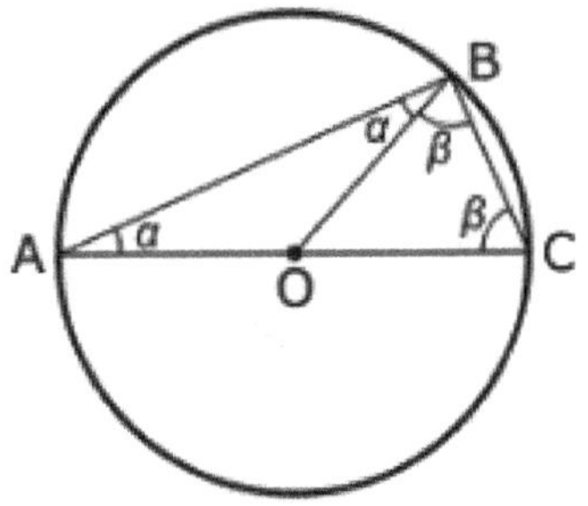

Thales Theorem

Thales

It is said Thales, most widely regarded as the first philosopher in the Greek tradition, measured the height of the pyramids by their shadows at the moment when his own shadow was equal to his height. Thales was said to have had a sacrifice in celebration of discovering Thales' Theorem just as Pythagoras had the Pythagorean Theorem.

Thales is the first known individual to use deductive reasoning applied to geometry, by deriving four corollaries to his theorem, and the first known individual to whom a mathematical discovery has been attributed. Indian and Babylonian mathematicians knew his theorem for special cases before he proved it. It is believed that Thales learned that an angle inscribed in a semicircle is a right angle during his travels to Babylon.

Pythagoras

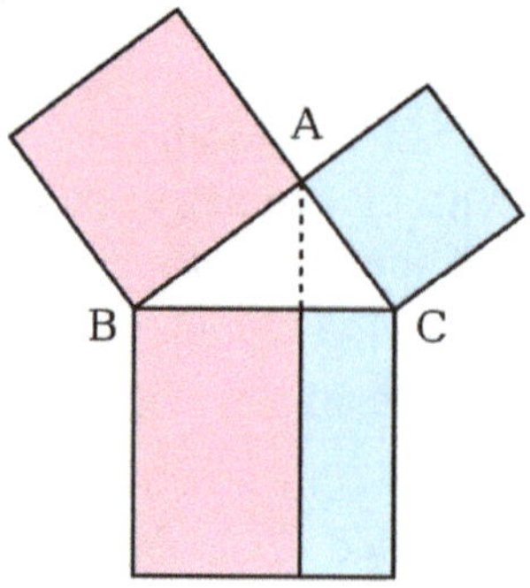

Proof of the Pythagorean Theorem in Euclid's *Elements*

Before 520 BC, on one of his visits to Egypt or Greece, Pythagoras might have met the c. 54 years older Thales. The systematic study of proof seems to have begun with the school of Pythagoras (i. e. the Pythagoreans) in the late sixth century BC. Indeed, the Pythagoreans, believing all was number, are the first philosophers to emphasize *form* rather than *matter*.

Heraclitus and Parmenides

The writing of Heraclitus (c. 535 – c. 475 BC) was the first place where the word *logos* was given special attention in ancient Greek philosophy,Heraclitus held that everything changes and all was fire and conflicting opposites, seemingly unified only by this *Logos*. He is known for his obscure sayings.

This *logos* holds always but humans always prove unable to understand it, both before hearing it and when they have first heard it. For though all things come to be in accordance with this *logos*, humans are like the inexperienced when they experience such words and deeds as I set out, distinguishing each in accordance with its nature and saying how it is. But other people fail to notice what they do when awake, just as they forget what they do while asleep.

—Diels-Kranz, 22B1

Parmenides has been called the discoverer of logic.

In contrast to Heraclitus, Parmenides held that all is one and nothing changes. He may have been a dissident Pythagorean, disagreeing that One (a number) produced the many. “X is not” must always be false or meaningless. What exists can in no way not exist. Our sense perceptions with its noticing of generation and destruction are in grievous error. Instead of sense perception, Parmenides advocated *logos* as the means to Truth. He has been called the discoverer of logic,

For this view, that That Which Is Not exists, can never predominate. You must debar your thought from this way of search, nor let ordinary experience in its variety force you along this way, (namely, that of allowing) the eye, sightless as it is, and the ear, full of sound, and the tongue, to rule; but (you must) judge by means of the Reason (Logos) the much-contested proof which is expounded by me. (B 7.1–8.2)

Zeno of Elea, a pupil of Parmenides, had the idea of a standard argument pattern found in the method of proof known as *reductio ad absurdum*. This is the technique of drawing an obviously false (that is, “absurd”) conclusion from an assumption, thus demonstrating that the assumption is false. Therefore, Zeno and his teacher are seen as the first to apply the art of logic. Plato’s dialogue Parmenides portrays Zeno as claiming to have written a book defending the monism of Parmenides by demonstrating the absurd consequence of assuming that there is plurality. Zeno famously used this method to develop his paradoxes in his arguments against motion. Such *dialectic* reasoning later became popular. The members of this school were called “dialecticians” (from a Greek word meaning “to discuss”).

Plato

Let no one ignorant of geometry enter here.

—Inscribed over the entrance to Plato’s Academy.

Plato’s academy

None of the surviving works of the great fourth-century philosopher Plato (428–347 BC) include any formal logic, but they include important contributions to the field of philosophical logic. Plato raises three questions:

- What is it that can properly be called true or false?
- What is the nature of the connection between the assumptions of a valid argument and its conclusion?
- What is the nature of definition?

The first question arises in the dialogue *Theaetetus*, where Plato identifies thought or opinion with talk or discourse (*logos*). The second question is a result of Plato's theory of Forms. Forms are not things in the ordinary sense, nor strictly ideas in the mind, but they correspond to what philosophers later called universals, namely an abstract entity common to each set of things that have the same name. In both the *Republic* and the *Sophist*, Plato suggests that the necessary connection between the assumptions of a valid argument and its conclusion corresponds to a necessary connection between "forms". The third question is about definition. Many of Plato's dialogues concern the search for a definition of some important concept (justice, truth, the Good), and it is likely that Plato was impressed by the importance of definition in mathematics. What underlies every definition is a Platonic Form, the common nature present in different particular things. Thus, a definition reflects the ultimate object of understanding, and is the foundation of all valid inference. This had a great influence on Plato's student Aristotle, in particular Aristotle's notion of the essence of a thing.

Aristotle

The logic of Aristotle, and particularly his theory of the syllogism, has had an enormous influence in Western thought. His logical works, called the *Organon*, are the earliest formal study of logic that have come down to modern times. Though it is difficult to determine the dates, the probable order of writing of Aristotle's logical works is:

Aristotle

- *The Categories*, a study of the ten kinds of primitive term.
- *The Topics* (with an appendix called *On Sophistical Refutations*), a discussion of dialectics.
- *On Interpretation*, an analysis of simple categorical propositions into simple terms, negation, and signs of quantity. It also contains a comprehensive treatment of the notions of opposition and conversion; chapter 7 is at the origin of the square of opposition (or logical square); chapter 9 contains the beginning of modal logic.
- *The Prior Analytics*, a formal analysis of what makes a syllogism (a valid argument, according to Aristotle).
- *The Posterior Analytics*, a study of scientific demonstration, containing Aristotle's mature views on logic.

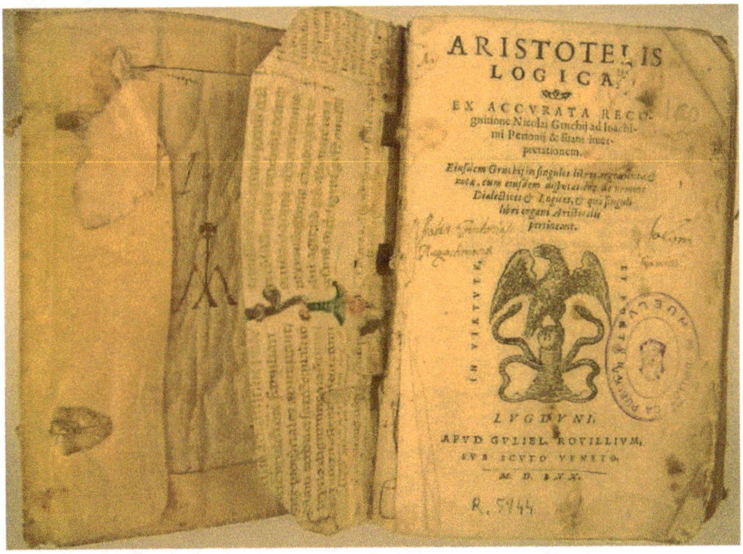

Aristotle's logic was still influential in the Renaissance

These works are of outstanding importance in the history of logic. Aristotle was the first logician to attempt a systematic analysis of logical syntax, of noun (or *term*), and of verb. He was the first *formal logician*, in that he demonstrated the principles of reasoning by employing variables to show the underlying logical form of an argument. He was looking for relations of dependence which characterize necessary inference, and distinguished the validity of these relations, from the truth of the premises (the soundness of the argument). He was the first to deal with the principles of contradiction and excluded middle in a systematic way.

In the *Categories*, he attempts to discern all the possible things to which a term can refer; this idea underpins his philosophical work *Metaphysics*, which itself had a profound influence on Western thought. The *Prior Analytics* contains his exposition of the "syllogism", where three important principles are applied for the first time in history: the use of variables, a purely formal treatment, and the use of an axiomatic system. He also developed a theory of non-formal logic (*i.e.,* the theory of fallacies), which is presented in *Topics* and *Sophistical Refutations*.

Stoics

The other great school of Greek logic is that of the Stoics. Stoic logic traces its roots back to the late 5th century BC philosopher Euclid of Megara, a pupil of Socrates and slightly older contemporary of Plato, probably following in the tradition of Parmenides and Zeno. His pupils and successors were called "Megarians", or "Eristics", and later the "Dialecticians". The two most important dialecticians of the Megarian school were Diodorus Cronus and Philo, who were active in the late 4th century BC. The Stoics adopted the Megarian logic and systemized it. The most important member of the school was Chrysippus (c. 278–c. 206 BC), who was its third head, and who formalized much of Stoic doctrine. He is supposed to have written over 700 works, including at least 300 on logic, almost none of which survive. Unlike with Aristotle, we have no complete works by the Megarians or the early Stoics, and have to rely mostly on accounts (sometimes hostile) by later sources, including prominently Diogenes Laertius, Sextus Empiricus, Galen, Aulus Gellius, Alexander of Aphrodisias, and Cicero.

Chrysippus of Soli

Three significant contributions of the Stoic school were (i) their account of modality, (ii) their theory of the Material conditional, and (iii) their account of meaning and truth.

- *Modality*. According to Aristotle, the Megarians of his day claimed there was no distinction between potentiality and actuality. Diodorus Cronus defined the possible as that which either is or will be, the impossible as what will not be true, and the contingent as that which either is already, or will be false. Diodorus is also famous for what is known as his Master argument, which states that each pair of the following 3 propositions contradicts the third proposition:

- Everything that is past is true and necessary.
- The impossible does not follow from the possible.
- What neither is nor will be is possible.

Diodorus used the plausibility of the first two to prove that nothing is possible if it neither is nor will be true. Chrysippus, by contrast, denied the second premise and said that the impossible could follow from the possible.

- *Conditional statements*. The first logicians to debate conditional statements were Diodorus and his pupil Philo of Megara. Sextus Empiricus refers three times to a debate between Diodorus and Philo. Philo regarded a conditional as true unless it has both a true antecedent and a false consequent. Precisely, let T_0 and T_1 be true statements, and let F_0 and F_1 be false statements; then, according to Philo, each of the following conditionals is a true statement, because it is not the case that the consequent is false while the antecedent is true (it is not the case that a false statement is asserted to follow from a true statement):
- If T_0, then T_1
- If F_0, then T_0
- If F_0, then F_1

The following conditional does not meet this requirement, and is therefore a false statement according to Philo:

- If T_0, *then* F_0

Indeed, Sextus says "According to Philo, there are three ways in which a conditional may be true, and one in which it may be false." Philo's criterion of truth is what would now be called a truth-functional definition of "if ... then"; it is the definition used in modern logic.

In contrast, Diodorus allowed the validity of conditionals only when the antecedent clause could never lead to an untrue conclusion. A century later, the Stoic philosopher Chrysippus attacked the assumptions of both Philo and Diodorus.

- *Meaning and truth*. The most important and striking difference between Megarian-Stoic logic and Aristotelian logic is that Megarian-Stoic logic concerns propositions, not terms, and is thus closer to modern propositional logic. The Stoics distinguished between utterance (*phone*), which may be noise, speech (*lexis*), which is articulate but which may be meaningless, and discourse (*logos*), which is meaningful utterance. The most original part of their theory is the idea that what is expressed by a sentence, called a *lekton*, is something real; this corresponds to what is now called a *proposition*. Sextus says that according to the Stoics, three things are linked together: that which signifies, that which is signified, and the object; for example, that which signifies is the word *Dion*, and that which is signified is what Greeks understand but barbarians do not, and the object is Dion himself.

MEDIEVAL LOGIC

Logic in the Middle East

A text by Avicenna, founder of Avicennian logic

The works of Al-Kindi, Al-Farabi, Avicenna, Al-Ghazali, Averroes and other Muslim logicians were based on Aristotelian logic and were important in communicating the ideas of the ancient world to the medieval West. Al-Farabi (Alfarabi) (873–950) was an Aristotelian logician who discussed the topics of future contingents, the number and relation of the categories, the relation between logic and grammar, and non-Aristotelian forms of inference. Al-Farabi also considered the theories of conditional syllogisms and analogical inference, which were part of the Stoic tradition of logic rather than the Aristotelian.

Ibn Sina (Avicenna) (980–1037) was the founder of Avicennian logic, which replaced Aristotelian logic as the dominant system of logic in the Islamic world, and also had an important influence on Western medieval writers such as Albertus Magnus. Avicenna wrote on the hypothetical syllogism and on the propositional calculus, which were both part of the Stoic logical tradition. He developed an original "temporally modalized" syllogistic theory, involving temporal logic and modal logic. He also made use of inductive logic, such as the methods of agreement, difference, and concomitant variation which are critical to the scientific method. One of Avicenna's ideas had a particularly important influence on Western logicians such as William of Ockham: Avicenna's word for a meaning or notion (*ma'na*), was translated by the scholastic logicians as the Latin *intentio*; in medieval logic and epistemology, this is a sign in the mind that naturally represents a thing. This was crucial to the development of Ockham's conceptualism: A universal term (*e.g.,* "man") does not signify a thing existing in reality, but rather a sign in the mind (*intentio in intellectu*) which represents many things in reality; Ockham cites Avicenna's commentary on *Metaphysics* V in support of this view.

Fakhr al-Din al-Razi (b. 1149) criticised Aristotle's "first figure" and formulated an early system of inductive logic, foreshadowing the system of inductive logic developed by John Stuart Mill (1806–1873). Al-Razi's work was seen by later Islamic scholars as marking a new direction for Islamic logic, towards a Post-Avicennian logic. This was further elaborated by his student Afdaladdîn al-Khûnajî (d. 1249), who developed a form of logic revolving around the subject matter of conceptions and assents. In response to this tradition, Nasir al-Din al-Tusi (1201–1274) began a tradition of Neo-Avicennian logic which remained faithful to Avicenna's work and existed as an alternative to the more dominant Post-Avicennian school over the following centuries.

The Illuminationist school was founded by Shahab al-Din Suhrawardi (1155–1191), who developed the idea of "decisive necessity", which refers to the reduction of all modalities (necessity, possibility, contingency and impossibility) to the single mode of necessity. Ibn al-Nafis (1213–1288) wrote a book on Avicennian logic, which was a commentary of Avicenna's *Al-Isharat* (*The Signs*) and *Al-Hidayah* (*The Guidance*). Ibn Taymiyyah (1263–1328), wrote the *Ar-Radd ‹ala al-Mantiqiyyin*, where he argued against the usefulness, though not the validity, of the syllogism and in favour of inductive reasoning. Ibn Taymiyyah also argued against the certainty of syllogistic arguments and in favour of analogy; his argument is that concepts founded on induction are themselves not certain but only probable, and thus a syllogism based on such concepts is no more certain than an argument based on analogy. He further claimed that induction itself is founded on a process of analogy. His model of analogical reasoning was based on that of juridical arguments. This model of analogy has been used in the recent work of John F. Sowa.

The *Sharh al-takmil fi 'l-mantiq* written by Muhammad ibn Fayd Allah ibn Muhammad Amin al-Sharwani in the 15th century is the last major Arabic work on logic that has been studied.However, "thousands upon thousands of pages" on logic were written between the 14th and 19th centuries, though only a fraction of the texts written during this period have been studied by historians, hence little is known about the original work on Islamic logic produced during this later period.

Logic in Medieval Europe

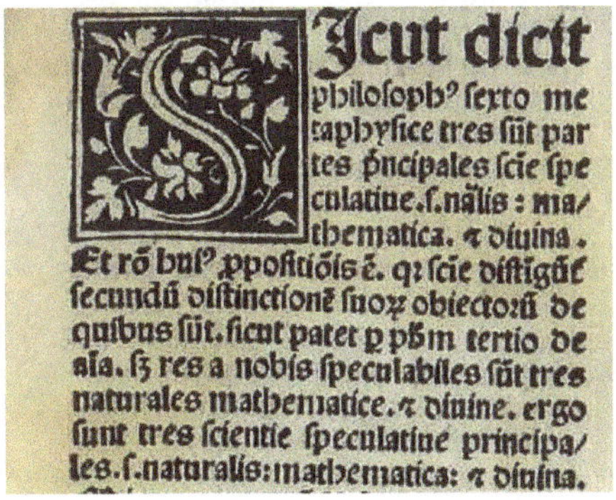
Sicut dicit philosoph⁹ sexto me taphysice tres sũt par tes pncipales scie spe culatiue.s.nãlis: ma- thematica. ⁊ diuina. Et rõ hui⁹ ppositiõis ẽ. q: scie distigũt secundũ distinctionẽ suoꝝ obiectoꝝ de quibus sũt. sicut patet p pħm tertio de aĩa. s; res a nobis speculabiles sũt tres naturales mathematice. ⁊ diuine. ergo sunt tres scientie speculatiue principa- les. s. naturalis: mathematica: ⁊ diuina.

Brito's questions on the *Old Logic*

"Medieval logic" (also known as "Scholastic logic") generally means the form of Aristotelian logic developed in medieval Europe throughout roughly the period 1200–1600. For centuries after Stoic logic had been formulated, it was the dominant system of logic in the classical world. When the study of logic resumed after the Dark Ages, the main source was the work of the Christian philosopher Boethius, who was familiar with some of Aristotle's logic, but almost none of the work of the Stoics. Until the twelfth century, the only works of Aristotle available in the West were the *Categories*, *On Interpretation*, and Boethius's translation of the Isagoge of Porphyry (a commentary on the Categories). These works were known as the "Old Logic" (*Logica Vetus* or *Ars Vetus*). An important work in this tradition was the *Logica Ingredientibus* of Peter Abelard (1079–1142). His direct influence was small, but his influence through pupils such as John of Salisbury was great, and his method of applying rigorous logical analysis to theology shaped the way that theological criticism developed in the period that followed.

By the early thirteenth century, the remaining works of Aristotle's *Organon* (including the *Prior Analytics*, *Posterior Analytics*, and the *Sophistical Refutations*) had been recovered in the West. Logical work until then was mostly paraphrasis or commentary on the work of Aristotle. The period from the middle of the thirteenth to the middle of the fourteenth century was one of significant developments in logic, particularly in three areas which were original, with little foundation in the Aristotelian tradition that came before. These were:

- The theory of supposition. Supposition theory deals with the way that predicates (*e.g.*, 'man') range over a domain of individuals (*e.g.*, all men). In the proposition 'every man is an animal', does the term 'man' range over or 'supposit for' men existing just in the present, or does the range include past and future men? Can a term supposit for a non-existing individual? Some medievalists have argued that this idea is a precursor of modern first-order logic. "The theory of supposition with the associated theories of *copulatio* (sign-capacity of adjectival terms), *ampliatio* (widening of referential domain), and *distributio* constitute one of the most original achievements of Western medieval logic".
- The theory of syncategoremata. Syncategoremata are terms which are necessary for logic, but which, unlike *categorematic* terms, do not signify on their own behalf, but 'co-signify' with other words. Examples of syncategoremata are ‹and', ‹not', ‹every', ‹if', and so on.
- The theory of consequences. A consequence is a hypothetical, conditional proposition: two propositions joined by the terms ‹if ... then'. For example, ‹if a man runs, then God exists' (*Si homo currit, Deus est*). A fully developed theory of consequences is given in Book III of William of Ockham's work Summa Logicae. There, Ockham distinguishes between 'material' and 'formal' consequences, which are roughly equivalent to the modern material implication and logical implication respectively. Similar accounts are given by Jean Buridan and Albert of Saxony.

The last great works in this tradition are the *Logic* of John Poinsot (1589–1644, known as John of St Thomas), the *Metaphysical Disputations* of Francisco Suarez (1548–1617), and the *Logica Demonstrativa* of Giovanni Girolamo Saccheri (1667–1733).

TRADITIONAL LOGIC

The Textbook Tradition

THE ARTES OF LOGIKE AND Rethorike, plainlie set foorth in the English tounge, easie to be learned and practised : togither with examples for the practise of the same for Methode, in the gouernement of the familie, prescribed in the word of God; And for the whole in the resolution or opening of certayne partes of Scripture, according to the same.

1584

Dudley Fenner's *Art of Logic* (1584)

Traditional logic generally means the textbook tradition that begins with Antoine Arnauld's and Pierre Nicole's *Logic, or the Art of Thinking*, better known as the *Port-Royal Logic*. Published in 1662, it was the most influential work on logic after Aristotle until the nineteenth century. The book presents a loosely Cartesian doctrine (that the proposition is a combining of ideas rather than terms, for example) within a framework that is broadly derived from Aristotelian and medieval term logic. Between 1664 and 1700, there were eight editions, and the book had considerable influence after that. The Port-Royal introduces the concepts of extension and intension. The account of propositions that Locke gives in the *Essay* is essentially that of the Port-Royal: "Verbal propositions, which are words, are the signs of our ideas, put together or separated in affirmative or negative sentences. So that proposition consists in the putting together or separating these signs, according as the things which they stand for agree or disagree."

Dudley Fenner helped popularize Ramist logic, a reaction against Aristotle. Another influential work was the *Novum Organum* by Francis Bacon, published in 1620. The title translates as "new instrument". This is a reference to Aristotle's work known as the *Organon*. In this work, Bacon rejects the syllogistic method of Aristotle in favor of an alternative procedure "which by slow and faithful toil gathers information from things and brings it into understanding". This method is known as inductive reasoning, a method which starts from empirical observation and proceeds to lower axioms or propositions; from these lower axioms, more general ones can be induced. For example, in finding the cause of a *phenomenal nature* such as heat, 3 lists should be constructed:

- The presence list: a list of every situation where heat is found.
- The absence list: a list of every situation that is similar to at least one of those of the presence list, except for the lack of heat.
- The variability list: a list of every situation where heat can vary.

Then, the *form nature* (or cause) of heat may be defined as that which is common to every situation of the presence list, and which is lacking from every situation of the absence list, and which varies by degree in every situation of the variability list.

Other works in the textbook tradition include Isaac Watts's *Logick: Or, the Right Use of Reason* (1725), Richard Whately's *Logic* (1826), and John Stuart Mill's *A System of Logic* (1843). Although the latter was one of the last great works in the tradition, Mill's view that the foundations of logic lie in introspection influenced the view that logic is best understood as a branch of psychology, a view which dominated the next fifty years of its development, especially in Germany.

Logic in Hegel's Philosophy

Georg Wilhelm Friedrich Hegel

G.W.F. Hegel indicated the importance of logic to his philosophical system when he condensed his extensive *Science of Logic* into a shorter work published in 1817 as the first volume of his *Encyclopaedia of the Philosophical Sciences.* The "Shorter" or "Encyclopaedia" *Logic*, as it is often known, lays out a series of transitions which leads from the most empty and abstract of categories—Hegel begins with "Pure Being" and

"Pure Nothing"—to the "Absolute, the category which contains and resolves all the categories which preceded it. Despite the title, Hegel's *Logic* is not really a contribution to the science of valid inference. Rather than deriving conclusions about concepts through valid inference from premises, Hegel seeks to show that thinking about one concept compels thinking about another concept (one cannot, he argues, possess the concept of "Quality" without the concept of "Quantity"); this compulsion is, supposedly, not a matter of individual psychology, because it arises almost organically from the content of the concepts themselves. His purpose is to show the rational structure of the "Absolute"—indeed of rationality itself. The method by which thought is driven from one concept to its contrary, and then to further concepts, is known as the Hegelian dialectic.

Although Hegel's *Logic* has had little impact on mainstream logical studies, its influence can be seen elsewhere:

- Carl von Prantl's *Geschichte der Logik in Abendland* (1855–1867).
- The work of the British Idealists, such as F.H. Bradley's *Principles of Logic* (1883).
- The economic, political, and philosophical studies of Karl Marx, and in the various schools of Marxism.

Logic and Psychology

Between the work of Mill and Frege stretched half a century during which logic was widely treated as a descriptive science, an empirical study of the structure of reasoning, and thus essentially as a branch of psychology. The German psychologist Wilhelm Wundt, for example, discussed deriving "the logical from the psychological laws of thought", emphasizing that "psychological thinking is always the more comprehensive form of thinking." This view was widespread among German philosophers of the period:

- Theodor Lipps described logic as "a specific discipline of psychology".
- Christoph von Sigwart understood logical necessity as grounded in the individual's compulsion to think in a certain way.
- Benno Erdmann argued that "logical laws only hold within the limits of our thinking".

Such was the dominant view of logic in the years following Mill's work. This psychological approach to logic was rejected by Gottlob Frege. It was also subjected to an extended and destructive critique by Edmund Husserl in the first volume of his *Logical Investigations* (1900), an assault which has been described as "overwhelming". Husserl argued forcefully that grounding logic in psychological observations implied that all logical truths remained unproven, and that skepticism and relativism were unavoidable consequences.

Such criticisms did not immediately extirpate what is called "psychologism". For example, the American philosopher Josiah Royce, while acknowledging the force of Husserl's critique, remained "unable to doubt" that progress in psychology would be accompanied by progress in logic, and vice versa.

Rise of Modern Logic

The period between the fourteenth century and the beginning of the nineteenth century had been largely one of decline and neglect, and is generally regarded as barren by historians of logic. The revival of logic occurred in the mid-nineteenth century, at the beginning of a revolutionary period where the subject developed into a rigorous and formalistic discipline whose exemplar was the exact method of proof used in mathematics. The development of the modern "symbolic" or "mathematical" logic during this period is the most significant in the 2000-year history of logic, and is arguably one of the most important and remarkable events in human intellectual history.

A number of features distinguish modern logic from the old Aristotelian or traditional logic, the most important of which are as follows: Modern logic is fundamentally a *calculus* whose rules of operation are determined only by the *shape* and not by the *meaning* of the symbols it employs, as in mathematics. Many logicians were impressed by the "success" of mathematics, in that there had been no prolonged dispute about any truly mathematical result. C.S. Peirce noted that even though a mistake in the evaluation of a definite integral by Laplace led to an error concerning the moon's orbit that persisted for nearly 50 years, the mistake, once spotted, was corrected without any serious dispute. Peirce contrasted this with the disputation and uncertainty surrounding traditional logic, and especially reasoning in metaphysics. He argued that a truly "exact" logic would depend upon mathematical, i.e., "diagrammatic" or "iconic" thought. "Those who follow such methods will ... escape all error except such as will be speedily corrected after it is once suspected". Modern logic is also "constructive" rather than "abstractive"; i.e., rather than abstracting and formalising theorems derived from ordinary language (or from psychological intuitions about validity), it constructs theorems by formal methods, then looks for an interpretation in ordinary language. It is entirely symbolic, meaning that even the logical constants (which the medieval logicians called "syncategoremata") and the categoric terms are expressed in symbols.

MODERN LOGIC

The development of modern logic falls into roughly five periods:

- The **embryonic period** from Leibniz to 1847, when the notion of a logical calculus was discussed and developed, particularly by Leibniz, but no schools were formed, and isolated periodic attempts were abandoned or went unnoticed.
- The **algebraic period** from Boole's Analysis to Schröder's *Vorlesungen*. In this period, there were more practitioners, and a greater continuity of development.
- The **logicist period** from the Begriffsschrift of Frege to the *Principia Mathematica* of Russell and Whitehead. The aim of the "logicist school" was to incorporate the logic of all mathematical and scientific discourse in a single unified system which, taking as a fundamental principle that all mathematical truths are logical, did not accept any non-logical terminology. The major logicists were Frege, Russell, and the early Wittgenstein. It culminates with the *Principia*, an important work which includes a thorough examination and attempted solution of the antinomies which had been an obstacle to earlier progress.

- The **metamathematical period** from 1910 to the 1930s, which saw the development of metalogic, in the finitist system of Hilbert, and the non-finitist system of Löwenheim and Skolem, the combination of logic and metalogic in the work of Gödel and Tarski. Gödel's incompleteness theorem of 1931 was one of the greatest achievements in the history of logic. Later in the 1930s, Gödel developed the notion of set-theoretic constructibility.
- The **period after World War II**, when mathematical logic branched into four inter-related but separate areas of research: model theory, proof theory, computability theory, and set theory, and its ideas and methods began to influence philosophy.

Embryonic Period

Leibniz

The idea that inference could be represented by a purely mechanical process is found as early as Raymond Llull, who proposed a (somewhat eccentric) method of drawing conclusions by a system of concentric rings. The work of logicians such as the Oxford Calculators led to a method of using letters instead of writing out logical calculations (*calculationes*) in words, a method used, for instance, in the *Logica magna* by Paul of Venice. Three hundred years after Llull, the English philosopher and logician Thomas Hobbes suggested that all logic and reasoning could be reduced to the mathematical operations of addition and subtraction. The same idea is found in the work of Leibniz, who had read both Llull and Hobbes, and who argued that logic can be represented through a combinatorial process or calculus. But, like Llull and Hobbes, he failed to develop a detailed or comprehensive system, and his work on this topic was not published until long after his death. Leibniz says that ordinary languages are subject to "countless ambiguities" and are unsuited for a calculus, whose task is to expose mistakes in inference arising from the forms and structures of words; hence, he proposed to identify an alphabet of human thought comprising fundamental concepts which could be composed to express complex ideas, and create a *calculus ratiocinator* that would make all arguments "as tangible as those of the Mathematicians, so that we can find our error at a glance, and when there are disputes among persons, we can simply say: Let us calculate."

Gergonne (1816) said that reasoning does not have to be about objects about which one has perfectly clear ideas, because algebraic operations can be carried out without having any idea of the meaning of the symbols involved. Bolzano anticipated a fundamental idea of modern proof theory when he defined logical consequence or "deducibility" in terms of variables:

> Hence I say that propositions $M, N, O,\ldots$ are *deducible* from propositions $A, B, C, D,\ldots$ with respect to variable parts $i, j,\ldots$, if every class of ideas whose substitution for $i, j,\ldots$ makes all of $A, B, C, D,\ldots$ true, also makes all of $M, N, O,\ldots$ true. Occasionally, since it is customary, I shall say that propositions $M, N, O,\ldots$ *follow*, or can be *inferred* or *derived*, from $A, B, C, D, \ldots$. Propositions $A, B, C, D,\ldots$ I shall call the *premises*, $M, N, O,\ldots$ the *conclusions*.

This is now known as semantic validity.

Algebraic Period

George Boole

Modern logic begins with what is known as the "algebraic school", originating with Boole and including Peirce, Jevons, Schröder, and Venn. Their objective was to develop a calculus to formalise reasoning in the area of classes, propositions, and probabilities. The school begins with Boole's seminal work *Mathematical Analysis of Logic* which appeared in 1847, although De Morgan (1847) is its immediate precursor. The fundamental idea of Boole's system is that algebraic formulae can be used to express logical relations. This idea occurred to Boole in his teenage years, working as an usher in a private school in Lincoln, Lincolnshire. For example, let x and y stand for classes let the symbol = signify that the classes have the same members, xy stand for the class containing all and only the members of x and y and so on. Boole calls these *elective symbols*, i.e. symbols which select certain objects for consideration. An expression in which elective symbols are used is called an *elective function*, and an equation of which the members are elective functions, is an *elective equation*. The theory of elective functions and their "development" is essentially the modern idea of truth-functions and their expression in disjunctive normal form.

Boole's system admits of two interpretations, in class logic, and propositional logic. Boole distinguished between "primary propositions" which are the subject of syllogistic theory, and "secondary propositions", which are the subject of propositional logic, and showed how under different "interpretations" the same algebraic system could represent both. An example of a primary proposition is "All inhabitants are either Europeans or Asiatics." An example of a secondary proposition is "Either all inhabitants are Europeans or they are all Asiatics." These are easily distinguished in modern propositional calculus, where it is also possible to show that the first follows from the second, but it is a significant disadvantage that there is no way of representing this in the Boolean system.

In his *Symbolic Logic* (1881), John Venn used diagrams of overlapping areas to express Boolean relations between classes or truth-conditions of propositions. In 1869 Jevons realised that Boole's methods could be mechanised, and constructed a "logical machine" which he showed to the Royal Society the following year. In 1885 Allan Marquand proposed an electrical version of the machine that is still extant (picture at the Firestone Library).

Charles Sanders Peirce

The defects in Boole's system (such as the use of the letter *v* for existential propositions) were all remedied by his followers. Jevons published *Pure Logic, or the Logic of Quality apart from Quantity* in 1864, where he suggested a symbol to signify exclusive or, which allowed Boole's system to be greatly simplified. This was usefully exploited by Schröder when he set out theorems in parallel columns in his *Vorlesungen* (1890–1905). Peirce (1880) showed how all the Boolean elective functions could be expressed by the use of a single primitive binary operation, "neither ... nor ..." and equally well "not both ... and ...", however, like many of Peirce's innovations, this remained unknown or unnoticed until Sheffer rediscovered it in 1913. Boole's early work also lacks the idea of the logical sum which originates in Peirce (1867), Schröder (1877) and Jevons (1890), and the concept of inclusion, first suggested by Gergonne (1816) and clearly articulated by Peirce (1870).

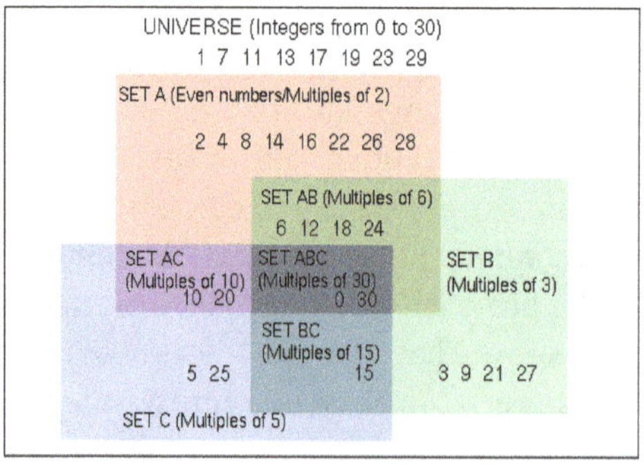

Boolean multiples

The success of Boole's algebraic system suggested that all logic must be capable of algebraic representation, and there were attempts to express a logic of relations in such form, of which the most ambitious was Schröder's monumental *Vorlesungen über die Algebra der Logik* ("Lectures on the Algebra of Logic", vol iii 1895), although the original idea was again anticipated by Peirce.

Boole's unwavering acceptance of Aristotle's logic is emphasized by the historian of logic John Corcoran in an accessible introduction to *Laws of Thought* Corcoran also wrote a point-by-point comparison of *Prior Analytics* and *Laws of Thought*. According to Corcoran, Boole fully accepted and endorsed Aristotle's logic. Boole's goals were "to go under, over, and beyond" Aristotle's logic by 1) providing it with mathematical foundations involving equations, 2) extending the class of problems it could treat — from assessing validity to solving equations — and 3) expanding the range of applications it could handle — e.g. from propositions having only two terms to those having arbitrarily many.

More specifically, Boole agreed with what Aristotle said; Boole's 'disagreements', if they might be called that, concern what Aristotle did not say. First, in the realm of foundations, Boole reduced the four propositional forms of Aristotelian logic to formulas in the form of equations — by itself a revolutionary idea. Second, in the realm of logic's problems, Boole's addition of equation solving to logic — another revolutionary idea — involved Boole's doctrine that Aristotle's rules of inference (the "perfect syllogisms") must be supplemented by rules for equation solving. Third, in the realm of applications, Boole's system could handle multi-term propositions and arguments whereas Aristotle could handle only two-termed subject-predicate propositions and arguments. For example, Aristotle's system could not deduce "No quadrangle that is a square is a rectangle that is a rhombus" from "No square that is a quadrangle is a rhombus that is a rectangle" or from "No rhombus that is a rectangle is a square that is a quadrangle".

Logicist Period

After Boole, the next great advances were made by the German mathematician Gottlob Frege. Frege's objective was the program of Logicism, i.e. demonstrating that arithmetic is identical with logic. Frege went much further than any of his predecessors in his rigorous and formal approach to logic, and his calculus or Begriffsschrift is important. Frege also tried to show that the concept of number can be defined by purely logical means, so that (if he was right) logic includes arithmetic and all branches of mathematics that are reducible to arithmetic. He was not the first writer to suggest this. In his pioneering work *Die Grundlagen der Arithmetik* (The Foundations of Arithmetic), sections 15–17, he acknowledges the efforts of Leibniz, J.S. Mill as well as Jevons, citing the latter's claim that "algebra is a highly developed logic, and number but logical discrimination."

Gottlob Frege.

Frege's first work, the *Begriffsschrift* ("concept script") is a rigorously axiomatised system of propositional logic, relying on just two connectives (negational and conditional), two rules of inference (*modus ponens* and substitution), and six axioms. Frege referred to the "completeness" of this system, but was unable to prove this. The most significant innovation, however, was his explanation of the quantifier in terms of mathematical functions. Traditional logic regards the sentence "Caesar is a man" as of fundamentally the same form as "all men are mortal." Sentences with a proper name subject were regarded as universal in character, interpretable as "every Caesar is a man". At the outset Frege abandons the traditional "concepts *subject* and *predicate*", replacing them with *argument* and *function* respectively, which he believes "will stand the test of time. It is easy to see how regarding a content as a function of an argument leads to the formation of concepts. Furthermore, the demonstration of the connection between the meanings of the words *if, and, not, or, there is, some, all,* and so forth, deserves attention". Frege argued that the quantifier expression "all men" does not have the same logical or semantic form as "all men", and that the universal proposition "every A is B" is a complex proposition involving two *functions*, namely ' – is A' and ' – is B' such that whatever satisfies the first, also satisfies the second. In modern notation, this would be expressed as

$$\forall\, x\big(A(x) \rightarrow B(x)\big)$$

In English, "for all x, if Ax then Bx". Thus only singular propositions are of subject-predicate form, and they are irreducibly singular, i.e. not reducible to a general proposition. Universal and particular propositions, by contrast, are not of simple subject-predicate form at all. If "all mammals" were the logical subject of the sentence "all mammals are land-dwellers", then to negate the whole sentence we would have to negate the predicate to give "all mammals are not land-dwellers". But this is not the case. This functional analysis of ordinary-language sentences later had a great impact on philosophy and linguistics.

This means that in Frege's calculus, Boole's "primary" propositions can be represented in a different way from "secondary" propositions. "All inhabitants are either men or women" is

$$\forall x\big(I(x) \to (M(x) \lor W(x))\big)$$

whereas "All the inhabitants are men or all the inhabitants are women" is

$$\forall x(I(x) \to M(x)) \lor \forall x(I(x) \to W(x))$$

As Frege remarked in a critique of Boole's calculus:

"The real difference is that I avoid the Boolean division into two parts ... and give a homogeneous presentation of the lot. In Boole the two parts run alongside one another, so that one is like the mirror image of the other, but for that very reason stands in no organic relation to it'. As well as providing a unified and comprehensive system of logic, Frege's calculus also resolved the ancient problem of multiple generality. The ambiguity of "every girl kissed a boy" is difficult to express in traditional logic, but Frege's logic resolves this through the different scope of the quantifiers. Thus

$$\forall x\big(G(x) \to \exists y(B(y) \land K(x,y))\big)$$

means that to every girl there corresponds some boy (any one will do) who the girl kissed. But

$$\exists x\big(B(x) \land \forall y(G(y) \to K(y,x))\big)$$

means that there is some particular boy whom every girl kissed. Without this device, the project of logicism would have been doubtful or impossible. Using it, Frege provided a definition of the ancestral relation, of the many-to-one relation, and of mathematical induction.

Ernst Zermelo

This period overlaps with the work of what is known as the "mathematical school", which included Dedekind, Pasch, Peano, Hilbert, Zermelo, Huntington, Veblen and Heyting. Their objective was the axiomatisation of branches of mathematics like geometry, arithmetic, analysis and set theory. Most notable was Hilbert's Program, which sought to ground all of mathematics to a finite set of axioms, proving its consistency by "finitistic" means and providing a procedure which would decide the truth or falsity of any mathematical statement. The standard axiomatization of the natural numbers is named the Peano axioms in his honor. Peano maintained a clear distinction between mathematical and logical symbols. While unaware of Frege's work, he independently recreated his logical apparatus based on the work of Boole and Schröder.

The logicist project received a near-fatal setback with the discovery of a paradox in 1901 by Bertrand Russell. This proved Frege's naive set theory led to a contradiction. Frege's theory contained the axiom that for any formal criterion, there is a set of all objects that meet the criterion. Russell showed that a set containing exactly the sets that are not members of themselves would contradict its own definition (if it is not a member of itself, it is a member of itself, and if it is a member of itself, it is not). This contradiction is now known as Russell's paradox. One important method of resolving this paradox was proposed by Ernst Zermelo. Zermelo set theory was the first axiomatic set theory. It was developed into the now-canonical Zermelo–Fraenkel set theory (ZF). Russell's paradox symbolically is as follows:

$$\textbf{Let } R = \{x \mid x \notin x\}, \textbf{ then } R \in R \iff R \notin R$$

The monumental Principia Mathematica, a three-volume work on the foundations of mathematics, written by Russell and Alfred North Whitehead and published 1910–13 also included an attempt to resolve the paradox, by means of an elaborate system of types: a set of elements is of a different type than is each of its elements (set is not the element; one element is not the set) and one cannot speak of the "set of all sets". The *Principia* was an attempt to derive all mathematical truths from a well-defined set of axioms and inference rules in symbolic logic.

Metamathematical Period

Kurt Gödel

The names of Gödel and Tarski dominate the 1930s, a crucial period in the development of metamathematics – the study of mathematics using mathematical methods to produce metatheories, or mathematical theories about other mathematical theories. Early investigations into metamathematics had been driven by Hilbert's program. Work on metamathematics culminated in the work of Gödel, who in 1929 showed that a given first-order sentence is deducible if and only if it is logically valid – i.e. it is true in every structure for its language. This is known as Gödel's completeness theorem. A year later, he proved two important theorems, which showed Hibert's program to be unattainable in its original form. The first is that no consistent system of axioms whose theorems can be listed by an effective procedure such as an algorithm or computer program is capable of proving all facts about the natural numbers. For any such system, there will always be statements about the natural numbers that are true, but that are unprovable within the system. The second is that if such a system is also capable of proving certain basic facts about the natural numbers, then the system cannot prove the consistency of the system itself. These two results are known as Gödel's incompleteness theorems, or simply *Gödel's Theorem*. Later in the decade, Gödel developed the concept of set-theoretic constructibility, as part of his proof that the axiom of choice and the continuum hypothesis are consistent with Zermelo–Fraenkel set theory. In proof theory, Gerhard Gentzen developed natural deduction and the sequent calculus. The former attempts to model logical reasoning as it 'naturally' occurs in practice and is most easily applied to intuitionistic logic, while the latter was devised to clarify the derivation of logical proofs in any formal system. Since Gentzen's work, natural deduction and sequent calculi have been widely applied in the fields of proof theory, mathematical logic and computer science. Gentzen also proved normalization and cut-elimination theorems for intuitionistic and classical logic which could be used to reduce logical proofs to a normal form.

Alfred Tarski

Alfred Tarski, a pupil of Łukasiewicz, is best known for his definition of truth and logical consequence, and the semantic concept of logical satisfaction. In 1933, he published (in Polish) *The concept of truth in formalized languages*, in which he proposed his semantic theory of truth: a sentence such as "snow is white" is true if and only if snow is white. Tarski's theory separated the metalanguage, which makes the statement about truth, from the object language, which contains the sentence whose truth is being asserted, and gave a correspondence (the T-schema) between phrases in the object language and elements of an interpretation. Tarski's approach to the difficult idea of explaining truth has been enduringly influential in logic and philosophy, especially in the development of model theory. Tarski also produced important work on the methodology of deductive systems, and on fundamental principles such as completeness, decidability, consistency and definability. According to Anita Feferman, Tarski "changed the face of logic in the twentieth century".

Alonzo Church and Alan Turing proposed formal models of computability, giving independent negative solutions to Hilbert's *Entscheidungsproblem* in 1936 and 1937, respectively. The *Entscheidungsproblem* asked for a procedure that, given any formal mathematical statement, would algorithmically determine whether the statement is true. Church and Turing proved there is no such procedure; Turing's paper introduced the halting problem as a key example of a mathematical problem without an algorithmic solution.

Church's system for computation developed into the modern λ-calculus, while the Turing machine became a standard model for a general-purpose computing device. It was soon shown that many other proposed models of computation were equivalent in power to those proposed by Church and Turing. These results led to the Church–Turing thesis that any deterministic algorithm that can be carried out by a human can be carried out by a Turing machine. Church proved additional undecidability results, showing that both Peano arithmetic and first-order logic are undecidable. Later work by Emil Post and Stephen Cole Kleene in the 1940s extended the scope of computability theory and introduced the concept of degrees of unsolvability.

The results of the first few decades of the twentieth century also had an impact upon analytic philosophy and philosophical logic, particularly from the 1950s onwards, in subjects such as modal logic, temporal logic, deontic logic, and relevance logic.

Logic after WWII

Saul Kripke

After World War II, mathematical logic branched into four inter-related but separate areas of research: model theory, proof theory, computability theory, and set theory.

In set theory, the method of forcing revolutionized the field by providing a robust method for constructing models and obtaining independence results. Paul Cohen introduced this method in 1963 to prove the independence of the continuum hypothesis and the axiom of choice from Zermelo–Fraenkel set theory. His technique, which was simplified and extended soon after its introduction, has since been applied to many other problems in all areas of mathematical logic.

Computability theory had its roots in the work of Turing, Church, Kleene, and Post in the 1930s and 40s. It developed into a study of abstract computability, which became known as recursion theory. The priority method, discovered independently by Albert Muchnik and Richard Friedberg in the 1950s, led to major advances in the understanding of the degrees of unsolvability and related structures. Research into higher-order computability theory demonstrated its connections to set theory. The fields of constructive analysis and computable analysis were developed to study the effective content of classical mathematical theorems; these in turn inspired the program of reverse mathematics. A separate branch of computability theory, computational complexity theory, was also characterized in logical terms as a result of investigations into descriptive complexity.

Model theory applies the methods of mathematical logic to study models of particular mathematical theories. Alfred Tarski published much pioneering work in the field, which is named after a series of papers he published under the title *Contributions to the theory of models*. In the 1960s, Abraham Robinson used model-theoretic techniques to develop calculus and analysis based on infinitesimals, a problem that first had been proposed by Leibniz.

In proof theory, the relationship between classical mathematics and intuitionistic mathematics was clarified via tools such as the realizability method invented by Georg Kreisel and Gödel's *Dialectica* interpretation. This work inspired the contemporary area of proof mining. The Curry-Howard correspondence emerged as a deep analogy between logic and computation, including a correspondence between systems of natural deduction and typed lambda calculi used in computer science. As a result, research into this class of formal systems began to address both logical and computational aspects; this area of research came to be known as modern type theory. Advances were also made in ordinal analysis and the study of independence results in arithmetic such as the Paris–Harrington theorem.

This was also a period, particularly in the 1950s and afterwards, when the ideas of mathematical logic begin to influence philosophical thinking. For example, tense logic is a formalised system for representing, and reasoning about, propositions qualified in terms of time. The philosopher Arthur Prior played a significant role in its development in the 1960s. Modal logics extend the scope of formal logic to include the elements of modality (for example, possibility and necessity). The ideas of Saul Kripke, particularly about possible worlds, and the formal system now called Kripke semantics have had a profound impact on analytic philosophy. His best known and most influential work is *Naming and Necessity* (1980). Deontic logics are closely related to modal logics: they attempt to capture the logical features of obligation, permission and related concepts. Although some basic novelties syncretizing mathematical and philosophical logic were shown by Bolzano in the early 1800s, it was Ernst Mally, a pupil of Alexius Meinong, who was to propose the first formal deontic system in his *Grundgesetze des Sollens*, based on the syntax of Whitehead's and Russell's propositional calculus.

Another logical system founded after World War II was fuzzy logic by Azerbaijani mathematician Lotfi Asker Zadeh in 1965.

LOGIC IN THE EAST

Logic in India

Logic began independently in ancient India and continued to develop to early modern times without any known influence from Greek logic. Medhatithi Gautama (c. 6th century BC) founded the *anviksiki* school of logic. The *Mahabharata* (12.173.45), around the 5th century BC, refers to the *anviksiki* and *tarka* schools of logic. Pāini (c. 5th century BC) developed a form of logic (to which Boolean logic has some similarities) for his formulation of Sanskrit grammar. Logic is described by Chanakya (c. 350-283 BC) in his *Arthashastra* as an independent field of inquiry.

Two of the six Indian schools of thought deal with logic: Nyaya and Vaisheshika. The Nyaya Sutras of Aksapada Gautama (c. 2nd century AD) constitute the core texts of the Nyaya school, one of the six orthodox schools of Hindu philosophy. This realist school developed a rigid five-member schema of inference involving an initial premise, a reason, an example, an application, and a conclusion. The idealist Buddhist philosophy became the chief opponent to the Naiyayikas. Nagarjuna (c. 150-250 AD), the founder of the Madhyamika ("Middle Way") developed an analysis known as the catukoi (Sanskrit), a "four-cornered" system of argumentation that involves the systematic examination and rejection of each of the 4 possibilities of a proposition, *P*:

- *P*; that is, being.
- not *P*; that is, not being.
- *P* and not *P*; that is, being and not being.
- not (*P* or not *P*); that is, neither being nor not being.

It is interesting to note that under propositional logic, De Morgan's laws imply that this is equivalent to the third case (*P* and not *P*), and is therefore superfluous; there are actually only 3 cases to consider.

However, Dignaga (c 480-540 AD) is sometimes said to have developed a formal syllogism, and it was through him and his successor, Dharmakirti, that Buddhist logic reached its height; it is contested whether their analysis actually constitutes a formal syllogistic system. In particular, their analysis centered on the definition of an inference-warranting relation, "vyapti", also known as invariable concomitance or pervasion. To this end, a doctrine known as "apoha" or differentiation was developed. This involved what might be called inclusion and exclusion of defining properties.

The difficulties involved in this enterprise, in part, stimulated the neo-scholastic school of Navya-Nyāya, which developed a formal analysis of inference in the sixteenth century. This later school began around eastern India and Bengal, and developed theories resembling modern logic, such as Gottlob Frege's "distinction between sense and reference of proper names" and his "definition of number," as well as the Navya-Nyaya theory of "restrictive conditions for universals" anticipating some of the developments in modern set theory. Since 1824, Indian logic attracted the attention of many Western scholars, and has had an influence on important 19th-century logicians such as Charles Babbage, Augustus De Morgan, and particularly George Boole, as confirmed by his wife Mary Everest Boole, who wrote in 1901 an "open letter to Dr Bose", which was titled "Indian Thought and Western Science in the Nineteenth Century" and stated: "Think what must have been the effect of the intense Hinduizing of three such men as Babbage, De Morgan and George Boole on the mathematical atmosphere of 1830-1865".

Dignāga's famous "wheel of reason" (*Hetucakra*) is a method of indicating when one thing (such as smoke) can be taken as an invariable sign of another thing (like fire), but the inference is often inductive and based on past observation. Matilal remarks that Dignāga's analysis is much like John Stuart Mill's Joint Method of Agreement and Difference, which is inductive.

In addition, the traditional five-member Indian syllogism, though deductively valid, has repetitions that are unnecessary to its logical validity. As a result, some commentators see the traditional Indian syllogism as a rhetorical form that is entirely natural in many cultures of the world, and yet not as a logical form—not in the sense that all logically unnecessary elements have been omitted for the sake of analysis.

Logic in China

In China, a contemporary of Confucius, Mozi, "Master Mo", is credited with founding the Mohist school, whose canons dealt with issues relating to valid inference and the conditions of correct conclusions. In particular, one of the schools that grew out of Mohism, the Logicians, are credited by some scholars for their early investigation of formal logic. Due to the harsh rule of Legalism in the subsequent Qin Dynasty, this line of investigation disappeared in China until the introduction of Indian philosophy by Buddhists.

REFERENCES

1. Boehner p. xiv
2. Oxford Companion p. 498; Bochenski, Part I Introduction, passim
3. Gottlob Frege. The Foundations of Arithmetic (PDF). p. 1.
4. Oxford Companion p. 500
5. Kneale, p. 2
6. Kneale p. 3
7. H. F. J. Horstmanshoff, Marten Stol, Cornelis Tilburg (2004), Magic and Rationality in Ancient Near Eastern and Graeco-Roman Medicine, p. 99, Brill Publishers, ISBN 90-04-13666-5.
8. D. Brown (2000), Mesopotamian Planetary Astronomy-Astrology , Styx Publications, ISBN 90-5693-036-2.
9. Heath, Mathematics in Aristotle, cited in Kneale, p. 5
10. Kneale, p. 16

11. Encyclopedia Britannica
12. Aristotle, Metaphysics Alpha, 983b18.
13. Smith, Sir William (1870). Dictionary of Greek and Roman biography and mythology. p. 1016.
14. Prof.T.Patronis & D.Patsopoulos The Theorem of Thales: A Study of the naming of theorems in school Geometry textbooks. Patras University. Retrieved 2012-02-12.
15. (Boyer 1991, "Ionia and the Pythagoreans" p. 43)
16. de Laet, Siegfried J. (1996). History of Humanity: Scientific and Cultural Development. UNESCO, Volume 3, p. 14. ISBN 92-3-102812-X
17. Boyer, Carl B. and Merzbach, Uta c. (2010). A History of Mathematics. John Wiley and Sons, Chapter IV. ISBN 0-470-63056-6
18. C. B. Boyer (1968)
19. Samuel Enoch Stumpf. Socrates to Sartre. p. 11.
20. F.E. Peters, Greek Philosophical Terms, New York University Press, 1967.
21. http://www.bard.edu/library/arendt/pdfs/Cornford-Parmenides.pdf
22. R. J. Hollingdale (1974). Western Philosophy: an introduction. p. 73.
23. http://www.wilbourhall.org/pdfs/From_religion_to_philosophy.pdf
24. Kneale p. 15
25. https://books.google.com/books?id=DPoqAAAAMAAJ&pg=PA170&lpg=PA170
26. Kneale p. 17
27. "forming an opinion is talking, and opinion is speech that is held not with someone else or aloud but in silence with oneself" Theaetetus 189E–190A
28. Kneale p. 20. For example, the proof given in the Meno that the square on the diagonal is double the area of the original square presumably involves the forms of the square and the triangle, and the necessary relation between them
29. Kneale p. 21
30. Zalta, Edward N. "Aristotle's Logic". Stanford University, 18 March 2000. Retrieved 13 March 2010.
31. See e.g. Aristotle's logic, Stanford Encyclopedia of Philosophy
32. Bochenski p. 63
33. "Throughout later antiquity two great schools of logic were distinguished, the Peripatetic which was derived from Aristotle, and the Stoic which was developed by Chrysippus from the teachings of the Megarians" – Kneale p. 113
34. Oxford Companion, article "Chrysippus", p. 134
35. Stanford Encyclopedia of Philosophy: Susanne Bobzien, Ancient Logic
36. K. Huelser, Die Fragmente zur Dialektik der Stoiker, 4 vols, Stuttgart 1986-7
37. Kneale 117–158
38. Metaphysics Eta 3, 1046b 29

39. Boethius, Commentary on the Perihermenias, Meiser p. 234
40. Epictetus, Dissertationes ed. Schenkel ii. 19. I.
41. Alexander p. 177
42. Sextus Empiricus, Adv. Math. viii, Section 113
43. Sextus Empiricus, Hypotyp. ii. 110, comp.
44. Cicero, Academica, ii. 47, de Fato, 6.
45. See e.g. Lukasiewicz p. 21
46. Sextus Bk viii., Sections 11, 12
47. See e.g. Routledge Encyclopedia of Philosophy Online Version 2.0, article 'Islamic philosophy'
48. History of logic: Arabic logic, Encyclopædia Britannica.
49. Feldman, Seymour (1964-11-26). "Rescher on Arabic Logic". The Journal of Philosophy. Journal of Philosophy, Inc. 61 (22): 724–734. doi:10.2307/2023632. ISSN 0022-362X. JSTOR 2023632. . Long, A. A.; D. N. Sedley (1987). The Hellenistic Philosophers. Vol 1: Translations of the principal sources with philosophical commentary. Cambridge: Cambridge University Press. ISBN 0-521-27556-3.
50. Dag Nikolaus Hasse (September 19, 2008). "Influence of Arabic and Islamic Philosophy on the Latin West". Stanford Encyclopedia of Philosophy. Retrieved 2009-10-13.
51. Richard F. Washell (1973), "Logic, Language, and Albert the Great", Journal of the History of Ideas 34 (3), pp. 445–450 .
52. Goodman, Lenn Evan (2003), Islamic Humanism, p. 155, Oxford University Press, ISBN 0-19-513580-6.
53. Goodman, Lenn Evan (1992); Avicenna, p. 188, Routledge, ISBN 0-415-01929-X.
54. Kneale p. 229
55. Kneale: p. 266; Ockham: Summa Logicae i. 14; Avicenna: Avicennae Opera Venice 1508 f87rb
56. Muhammad Iqbal, The Reconstruction of Religious Thought in Islam, "The Spirit of Muslim Culture" (cf. and)
57. Tony Street (July 23, 2008). "Arabic and Islamic Philosophy of Language and Logic". Stanford Encyclopedia of Philosophy. Retrieved 2008-12-05.
58. Dr. Lotfollah Nabavi, Sohrevardi's Theory of Decisive Necessity and kripke's QSS System, Journal of Faculty of Literature and Human Sciences.
59. Dr. Abu Shadi Al-Roubi (1982), "Ibn Al-Nafis as a philosopher", Symposium on Ibn al-Nafis, Second International Conference on Islamic Medicine: Islamic Medical Organization, Kuwait (cf.Ibn al-Nafis As a Philosopher, Encyclopedia of Islamic World).
60. See pp. 253–254 of Street, Tony (2005). "Logic". In Peter Adamson and Richard C. Taylor (edd.). The Cambridge Companion to Arabic Philosophy. Cambridge

University Press. pp. 247–265. ISBN 978-0-521-52069-0.

61. Ruth Mas (1998). "Qiyas: A Study in Islamic Logic" (PDF). Folia Orientalia. 34: 113–128. ISSN 0015-5675.
62. John F. Sowa; Arun K. Majumdar (2003). "Analogical reasoning". Conceptual Structures for Knowledge Creation and Communication, Proceedings of ICCS 2003. Berlin: Springer-Verlag., pp. 16-36
63. Nicholas Rescher and Arnold vander Nat, "The Arabic Theory of Temporal Modal Syllogistic", in George Fadlo Hourani (1975), Essays on Islamic Philosophy and Science, pp. 189–221, State University of New York Press, ISBN 0-87395-224-3.
64. Kneale p. 198
65. Stephen Dumont, article "Peter Abelard" in Gracia and Noone p. 492
66. Kneale, pp. 202–3
67. See e.g. Kneale p. 225
68. Boehner p. 1
69. Boehner pp. 19–76
70. Boehner p. 29
71. Boehner p. 30
72. Ebbesen 1981
73. Boehner pp. 54–5
74. Oxford Companion p. 504, article "Traditional logic"
75. Buroker xxiii
76. (Locke, An Essay Concerning Human Understanding, IV. 5. 6)
77. Farrington, 1964, 89
78. N. Abbagnano, "Psychologism" in P. Edwards (ed) The Encyclopaedia of Philosophy, MacMillan, 1967
79. Of the German literature in this period, Robert Adamson wrote "Logics swarm as bees in springtime..."; Robert Adamson, A Short History of Logic, Wm. Blackwood & Sons, 1911, page 242
80. Carl von Prantl (1855-1867), Geschichte von Logik in Abendland, Leipsig: S. Hirzl, anastatically reprinted in 1997, Hildesheim: Georg Olds.
81. See e.g. Psychologism, Stanford Encyclopedia of Philosophy
82. Wilhelm Wundt, Logik (1880–1883); quoted in Edmund Husserl, Logical Investigations, translated J.N. Findlay, Routledge, 2008, Volume 1, pp. 115–116.
83. Theodor Lipps, Grundzüge der Logik (1893); quoted in Edmund Husserl, Logical Investigations, translated J.N. Findlay, Routledge, 2008, Volume 1, p. 40
84. Christoph von Sigwart, Logik (1873–78); quoted in Edmund Husserl, Logical Investigations, translated J.N. Findlay, Routledge, 2008, Volume 1, p. 51
85. Benno Erdmann, Logik (1892); quoted in Edmund Husserl, Logical Investigations, translated J.N. Findlay, Routledge, 2008, Volume 1, p. 96

86. Dermot Moran, "Introduction"; Edmund Husserl, Logical Investigations, translated J.N. Findlay, Routledge, 2008, Volume 1, p. xxi
87. Michael Dummett, "Preface"; Edmund Husserl, Logical Investigations, translated J.N. Findlay, Routledge, 2008, Volume 1, p. xvii
88. Josiah Royce, "Recent Logical Enquiries and their Psychological Bearings" (1902) in John J. McDermott (ed) The Basic Writings of Josiah Royce Volume 2, Fordham University Press, 2005, p. 661
89. Bochenski, p. 266
90. Peirce 1896
91. See Bochenski p. 269
92. Oxford Companion p. 499
93. Edith Sylla (1999), "Oxford Calculators", in The Cambridge Dictionary of Philosophy, Cambridge, Cambridgeshire: Cambridge.
94. El. philos. sect. I de corp 1.1.2.
95. Bochenski p. 274
96. Rutherford, Donald, 1995, "Philosophy and language" in Jolley, N., ed., The Cambridge Companion to Leibniz. Cambridge Univ. Press.
97. Wiener, Philip, 1951. Leibniz: Selections. Scribner.
98. Essai de dialectique rationelle, 211n, quoted in Bochenski p. 277.
99. Bolzano, Bernard (1972). George, Rolf, ed. The Theory of Science: Die Wissenschaftslehre oder Versuch einer Neuen Darstellung der Logik. Translated by George Rolf. University of California Press. p. 209. ISBN 9780520017870.
100. See e.g. Bochenski p. 296 and passim
101. Before publishing, he wrote to De Morgan, who was just finishing his work Formal Logic. De Morgan suggested they should publish first, and thus the two books appeared at the same time, possibly even reaching the bookshops on the same day. cf. Kneale p. 404
102. Kneale p. 404
103. Kneale p. 407
104. Boole (1847) p. 16
105. Boole 1847 pp. 58–9
106. Beaney p. 11
107. Kneale p. 422
108. Peirce, "A Boolean Algebra with One Constant", 1880 MS, Collected Papers v. 4, paragraphs 12–20, reprinted Writings v. 4, pp. 218-21. Google Preview.
109. Trans. Amer. Math. Soc., xiv (1913), pp. 481–8. This is now known as the Sheffer stroke
110. Bochenski 296
111. See CP III

112. George Boole. 1854/2003. The Laws of Thought, facsimile of 1854 edition, with an introduction by J. Corcoran. Buffalo: Prometheus Books (2003). Reviewed by James van Evra in Philosophy in Review.24 (2004) 167–169.
113. JOHN CORCORAN, Aristotle's Prior Analytics and Boole's Laws of Thought, History and Philosophy of Logic, vol. 24 (2003), pp. 261–288.
114. Kneale p. 435
115. Jevons, The Principles of Science, London 1879, p. 156, quoted in Grundlagen 15
116. Beaney p. 10 – the completeness of Frege's system was eventually proved by Jan Łukasiewicz in 1934
117. See for example the argument by the medieval logician William of Ockham that singular propositions are universal, in Summa Logicae III. 8 (??)
118. Frege 1879 in van Heijenoort 1967, p. 7
119. "On concept and object" p. 198; Geach p. 48
120. BLC p. 14, quoted in Beaney p. 12
121. See e.g. The Internet Encyclopedia of Philosophy, article "Frege"
122. Van Heijenoort 1967, p. 83
123. See e.g. Potter 2004
124. Zermelo 1908
125. Feferman 1999 p. 1
126. Girard, Jean-Yves; Paul Taylor; Yves Lafont (1990) . Proofs and Types. Cambridge University Press (Cambridge Tracts in Theoretical Computer Science, 7). ISBN 0-521-37181-3.
127. Alex Sakharov. "Cut Elimination Theorem". MathWorld.
128. Feferman and Feferman 2004, p. 122, discussing "The Impact of Tarski's Theory of Truth".
129. Feferman 1999, p. 1
130. See e.g. Barwise, Handbook of Mathematical Logic
131. The Independence of the Continuum Hypothesis, II Paul J. Cohen Proceedings of the National Academy of Sciences of the United States of America, Vol. 51, No. 1. (Jan. 15, 1964), pp. 105-110.
132. Many of the foundational papers are collected in The Undecidable (1965) edited by Martin Davis
133. Jerry Fodor, "Water's water everywhere", London Review of Books, 21 October 2004
134. See Philosophical Analysis in the Twentieth Century: Volume 2: The Age of Meaning, Scott Soames: "Naming and Necessity is among the most important works ever, ranking with the classical work of Frege in the late nineteenth century, and of Russell, Tarski and Wittgenstein in the first half of the twentieth century". Cited in Byrne, Alex and Hall, Ned. 2004. ‹Necessary Truths'. Boston Review October/November 2004

135. Bochenski p. 446

136. S. C. Vidyabhusana (1971). A History of Indian Logic: Ancient, Mediaeval, and Modern Schools.

137. R. P. Kangle (1986). The Kautiliya Arthashastra (1.2.11). Motilal Banarsidass.

138. Bochenski p. 417 and passim

139. Bochenski pp. 431–7

140. Matilal, Bimal Krishna (1998). The Character of Logic in India. Albany, NY: State University of New York Press. pp. 12, 18. ISBN 9780791437407.

141. Bochenksi p. 441

142. Kisor Kumar Chakrabarti (June 1976). "Some Comparisons Between Frege's Logic and Navya-Nyaya Logic". Philosophy and Phenomenological Research. International Phenomenological Society. 36 (4): 554–563. doi:10.2307/2106873. JSTOR 2106873. This paper consists of three parts. The first part deals with Frege's distinction between sense and reference of proper names and a similar distinction in Navya-Nyaya logic. In the second part we have compared Frege's definition of number to the Navya-Nyaya definition of number. In the third part we have shown how the study of the so-called ‹restrictive conditions for universals' in Navya-Nyaya logic anticipated some of the developments of modern set theory.

143. Boole, Mary Everest "Collected Works" eds E M Cobham and E S Dummer London, Daniel 1931. Letter also published in the Ceylon National Review in 1909, and published as a separate pamphlet "The Psychologic Aspect of Imperialism" in 1911.

144. Jonardon Ganeri (2001). Indian logic: a reader. Routledge. p. vii. ISBN 0-7007-1306-9

145. Matilal, 17

CHAPTER 2

Mathematical Induction

Chapter Contents

The natural numbers are the counting numbers: 1, 2, 3, 4, etc. Mathematical induction is a technique for proving a statement -- a theorem, or a formula -- that is asserted about every natural number.

INTRODUCTION

Mathematical induction is a mathematical proof technique used to prove a given statement about any well-ordered set. Most commonly, it is used to establish statements for the set of all natural numbers.

Mathematical induction is a form of direct proof, usually done in two steps. When trying to prove a given statement for a set of natural numbers, the first step, known as the base case, is to prove the given statement for the first natural number. The second step, known as the inductive step, is to prove that, if the statement is assumed to be true for any one natural number, then it must be true for the next natural number as well. Having proved these two steps, the rule of inference establishes the statement to be true for all natural numbers. In common terminology, using the stated approach is referred to as using the Principle of mathematical induction.

The method can be extended to prove statements about more general well-founded structures, such as trees; this generalization, known as structural induction, is used in mathematical logic and computer science. Mathematical induction in this extended sense is closely related to recursion. Mathematical induction, in some form, is the foundation of all correctness proofs for computer programs.

Although its name may suggest otherwise, mathematical induction should not be misconstrued as a form of inductive reasoning (also see Problem of induction). Mathematical induction is an inference rule used in proofs. In mathematics, proofs including those using mathematical induction are examples of deductive reasoning, and inductive reasoning is excluded from proofs.

In 370 BC, Plato's Parmenides may have contained an early example of an implicit inductive proof. The earliest implicit traces of mathematical induction can be found in Euclid's proof that the number of primes is infinite and in Bhaskara's "cyclic method". An opposite iterated technique, counting *down* rather than up, is found in the Sorites paradox, where it was argued that if 1,000,000 grains of sand formed a heap, and removing one grain from a heap left it a heap, then a single grain of sand (or even no grains) forms a heap.

An implicit proof by mathematical induction for arithmetic sequences was introduced in the *al-Fakhri* written by al-Karaji around 1000 AD, who used it to prove the binomial theorem and properties of Pascal's triangle.

None of these ancient mathematicians, however, explicitly stated the inductive hypothesis. Another similar case (contrary to what Vacca has written, as Freudenthal carefully showed) was that of Francesco Maurolico in his *Arithmeticorum libri duo* (1575), who used the technique to prove that the sum of the first n odd integers is n^2. The first explicit formulation of the principle of induction was given by Pascal in his *Traité du triangle arithmétique* (1665). Another Frenchman, Fermat, made ample use of a related principle, indirect proof by infinite descent. The inductive hypothesis was also employed by the Swiss Jakob Bernoulli, and from then on it became more or less well known. The modern rigorous and systematic treatment of the principle came only in the 19th century, with George Boole, Augustus de Morgan, Charles Sanders Peirce,Giuseppe Peano, and Richard Dedekind.

Description

The simplest and most common form of mathematical induction infers that a statement involving a natural number n holds for all values of n. The proof consists of two steps:

- The **basis (base case)**: prove that the statement holds for the first natural number n. Usually, $n = 0$ or $n = 1$, rarely, $n = -1$ (although not a natural number, the extension of the natural numbers to -1 is still a well-ordered set).
- The **inductive step**: prove that, if the statement holds for some natural number n, then the statement holds for $n + 1$.

The hypothesis in the inductive step that the statement holds for some n is called the **induction hypothesis** (or **inductive hypothesis**). To perform the inductive step, one assumes the induction hypothesis and then uses this assumption to prove the statement for $n + 1$.

Whether $n = 0$ or $n = 1$ depends on the definition of the natural numbers. If 0 is considered a natural number, as is common in the fields of combinatorics and mathematical logic, the base case is given by $n = 0$. If, on the other hand, 1 is taken as the first natural number, then the base case is given by $n = 1$.

Example

Mathematical induction can be used to prove that the following statement, $P(n)$, holds for all natural numbers n.

$$0 + 1 + 2 + \cdots + n = \frac{n(n+1)}{2}.$$

$P(n)$ gives a formula for the sum of the natural numbers less than or equal to number n. The proof that $P(n)$ is true for each natural number n proceeds as follows.

Basis: Show that the statement holds for $n = 0.P(0)$ amounts to the statement:

$$0 = \frac{0 \cdot (0+1)}{2}.$$

In the left-hand side of the equation, the only term is 0, and so the left-hand side is simply equal to 0.In the right-hand side of the equation, $0 \cdot (0 + 1)/2 = 0$.The two sides are equal, so the statement is true for $n = 0$. Thus it has been shown that $P(0)$ holds.

Inductive step: Show that *if* $P(k)$ holds, then also $P(k + 1)$ holds. This can be done as follows.

Assume $P(k)$ holds (for some unspecified value of k). It must then be shown that $P(k + 1)$ holds, that is:

$$(0 + 1 + 2 + \cdots + k) + (k+1) = \frac{(k+1)((k+1)+1)}{2}.$$

Using the induction hypothesis that $P(k)$ holds, the left-hand side can be rewritten to:

$$\frac{k(k+1)}{2} + (k+1).$$

Algebraically:

$$\begin{aligned} \frac{k(k+1)}{2} + (k+1) &= \frac{k(k+1) + 2(k+1)}{2} \\ &= \frac{(k+1)(k+2)}{2} \\ &= \frac{(k+1)((k+1)+1)}{2} \end{aligned}$$

thereby showing that indeed $P(k + 1)$ holds.

Since both the basis and the inductive step have been performed, by mathematical induction, the statement $P(n)$ holds for all natural numbers n. Q.E.D.

Axiom of Induction

Mathematical induction as an inference rule can be formalized as a second-order axiom. The *axiom of induction* is, in logical symbols,

$$\forall P.\ [[P(0) \land \forall(k \in \mathbb{N}).\ [P(k) \Rightarrow P(k+1)]] \Rightarrow \forall(n \in \mathbb{N}).\ P(n)]$$

where P is any predicate and k and n are both natural numbers.

In words, the basis $P(0)$ and the inductive step (namely, that the inductive hypothesis $P(k)$ implies $P(k + 1)$) together imply that $P(n)$ for any natural number n. The axiom of induction asserts that the validity of inferring that $P(n)$ holds for any natural number n from the basis and the inductive step.

Note that the first quantifier in the axiom ranges over *predicates* rather than over individual numbers. This is a second-order quantifier, which means that this axiom is stated in second-order logic. Axiomatizing arithmetic induction in first-order logic requires an axiom schema containing a separate axiom for each possible predicate. The article Peano axioms contains further discussion of this issue.

Characterizing the Structure of N by the Induction Axiom

Having proven the base case and the inductive step, then the structure of {\displaystyle \mathbb {N} } is such that any value can be obtained by performing the inductive step repeatedly. It may be helpful to think of the domino effect. Consider a half line of dominoes each standing on end, and extending infinitely to the right (see picture). Suppose that:

- The first domino falls right.
- If a (fixed but arbitrary) domino falls right, then its righthand neighbor also falls right.

With these assumptions one can conclude (using mathematical induction) that all of the dominoes will fall right.

If the dominoes are arranged in another way, this conclusion needn't hold (see Peano axioms § Formulation for a counter example). Similarly, the induction axiom describes an essential property of N, viz. that each of its members can be reached from 0 by sufficiently often adding 1. While there is only one structure that satisfies all Peano axioms (including induction), there is no set of only first-order axioms that fulfils the same task.

Variants

In practice, proofs by induction are often structured differently, depending on the exact nature of the property to be proved.

Induction Basis other Than 0 or 1

If one wishes to prove a statement not for all natural numbers but only for all numbers greater than or equal to a certain number b, then the proof by induction consists of:

- Showing that the statement holds when $n = b$.
- Showing that if the statement holds for $n = m \geq b$ then the same statement also holds for $n = m + 1$.

This can be used, for example, to show that $n^2 \geq 3n$ for $n \geq 3$. A more substantial example is a proof that

$$\frac{n^n}{3^n} < n! < \frac{n^n}{2^n} \text{ for } n \geq 6.$$

In this way, one can prove that $P(n)$ holds for all $n \geq 1$, or even $n \geq -5$. This form of mathematical induction is actually a special case of the previous form because if the statement to be proved is $P(n)$ then proving it with these two rules is equivalent with proving $P(n + b)$ for all natural numbers n with the first two steps.

Induction Basis Equal To 2

In mathematics, many standard functions, including operations such as "+" and relations such as "=", are binary, meaning that they take two arguments. Often these functions possess properties that implicitly extend them to more than two arguments. For example, once addition $a + b$ is defined and is known to satisfy the associativity property $(a + b) + c = a + (b + c)$, then the ternary addition $a + b + c$ makes sense, either as $(a + b) + c$ or as $a + (b + c)$. Similarly, many axioms and theorems in mathematics are stated only for the binary versions of mathematical operations and relations, and implicitly extend to higher-arity versions.

Suppose that one wishes to prove a statement about an n-ary operation implicitly defined from a binary operation, using mathematical induction on n. In this case it is natural to take 2 for the induction basis.

Example: Product Rule for the Derivative

In this example, the binary operation in question is multiplication (of functions). The usual product rule for the derivative taught in calculus states:

$$(fg)' = f'g + g'f.$$

This can be generalized to a product of n functions. One has

$$(f_1 f_2 f_3 \cdots f_n)'$$
$$= (f_1' f_2 f_3 \cdots f_n) + (f_1 f_2' f_3 \cdots f_n) + (f_1 f_2 f_3' \cdots f_n) + \cdots + (f_1 f_2 \cdots f_{n-1} f_n').$$

or in logarithmic derivative form

$$(f_1 f_2 f_3 \cdots f_n)'/(f_1 f_2 f_3 \cdots f_n)$$
$$= (f_1'/f_1) + (f_2'/f_2) + (f_3'/f_3) + \cdots + (f_n'/f_n).$$

In each of the n terms of the usual form, just one of the factors is a derivative; the others are not.

When this general fact is proved by mathematical induction, the $n = 0$ case is trivial $(1)' = 0$ (since the empty product is 1, and the empty sum is 0). The $n = 1$ case is also trivial, $f_1' = f_1'$. And for each $n \geq 3$, the case is easy to prove from the preceding $n - 1$ case. The real difficulty lies in the $n = 2$ case, which is why that is the one stated in the standard product rule.

Induction on More than One Counter

It is sometimes desirable to prove a statement involving two natural numbers, n and m, by iterating the induction process. That is, one performs a basis step and an inductive step for n, and in each of those performs a basis step and an inductive step for m. See, for example, the proof of commutativity accompanying *addition of natural numbers*. More complicated arguments involving three or more counters are also possible.

Infinite Descent

The method of infinite descent was one of Pierre de Fermat's favorites. This method of proof can assume several slightly different forms. For example, it might begin by showing that if a statement is true for a natural number n it must also be true for some smaller natural number m ($m < n$). Using mathematical induction (implicitly) with the inductive hypothesis being that the statement is false for all natural numbers less than or equal to m, one may conclude that the statement cannot be true for any natural number n.

Although this particular form of infinite-descent proof is clearly a mathematical induction, whether one holds all proofs "by infinite descent" to be mathematical inductions depends on how one defines the term "proof by infinite descent." One might, for example, use the term to apply to proofs in which the well-ordering of the natural numbers is assumed, but not the principle of induction. Such, for example, is the usual proof that 2 has no rational square root (see Infinite descent).

Prefix Induction

The most common form of induction requires proving that

$$\forall k(P(k) \to P(k+1))$$

or equivalently

$$\forall k(P(k-1) \to P(k))$$

whereupon the induction principle "automates" n applications of this inference in getting from $P(0)$ to $P(n)$. This could be called "predecessor induction" because each step proves something about a number from something about that number's predecessor.

A variant of interest in computational complexity is "prefix induction", in which one needs to prove

$$\forall k(P(k) \to P(2k) \land P(2k+1))$$

or equivalently

$$\forall k\left(P\left(\left\lfloor\frac{k}{2}\right\rfloor\right) \to P(k)\right)$$

The induction principle then "automates" log n applications of this inference in getting from $P(0)$ to $P(n)$. (It is called "prefix induction" because each step proves something about a number from something about the "prefix" of that number formed by truncating the low bit of its binary representation.)

If traditional predecessor induction is interpreted computationally as an n-step loop, prefix induction corresponds to a log n-step loop, and thus proofs using prefix induction are "more feasibly constructive" than proofs using predecessor induction.

Predecessor induction can trivially simulate prefix induction on the same statement. Prefix induction can simulate predecessor induction, but only at the cost of making the statement more syntactically complex (adding a bounded universal quantifier), so the interesting results relating prefix induction to polynomial-time computation depend on excluding unbounded quantifiers entirely, and limiting the alternation of bounded universal and existential quantifiers allowed in the statement. See

One could take it a step farther to "prefix of prefix induction": one must prove

$$\forall k\left(P\left(\left\lfloor\sqrt{k}\right\rfloor\right) \to P(k)\right)$$

whereupon the induction principle "automates" log log n applications of this inference in getting from $P(0)$ to $P(n)$. This form of induction has been used, analogously, to study log-time parallel computation.

Complete Induction

Another variant, called complete induction, course of values induction or strong induction (in contrast to which the basic form of induction is sometimes known as weak induction) makes the inductive step easier to prove by using a stronger hypothesis: one proves the statement $P(m + 1)$ under the assumption that $P(n)$ holds for all natural n less than $m + 1$; by contrast, the basic form only assumes $P(m)$. The name "strong induction" does not mean that this method can prove more than "weak induction", but merely refers to the stronger hypothesis used in the inductive step; in fact the two methods are equivalent, as explained below. In this form of complete induction one still has to prove the base case, $P(0)$, and it may even be necessary to prove extra base cases such as $P(1)$ before the general argument applies, as in the example below of the Fibonacci number F_n.

Although the form just described requires one to prove the base case, this is unnecessary if one can prove P(m) (assuming P(n) for all lower n) for all $m \geq 0$. This is a special case of transfinite induction as described below. In this form the base case is subsumed by the case $m = 0$, where P(0) is proved with no other P(n) assumed; this case may need to be handled separately, but sometimes the same argument applies for $m = 0$ and $m > 0$, making the proof simpler and more elegant. In this method it is, however, vital to ensure that the proof of P(m) does not implicitly assume that $m > 0$, e.g. by saying "choose an arbitrary $n < m$" or assuming that a set of m elements has an element.

Complete induction is equivalent to ordinary mathematical induction as described above, in the sense that a proof by one method can be transformed into a proof by the other. Suppose there is a proof of P(n) by complete induction. Let Q(n) mean "P(m) holds for all m such that $0 \leq m \leq n$". Then Q(n) holds for all n if and only if P(n) holds for all n, and our proof of P(n) is easily transformed into a proof of Q(n) by (ordinary) induction. If, on the other hand, P(n) had been proven by ordinary induction, the proof would already effectively be one by complete induction: P(0) is proved in the base case, using no assumptions, and P(n + 1) is proved in the inductive step, in which one may assume all earlier cases but need only use the case P(n).

Examples of Complete Induction

Complete induction is most useful when several instances of the inductive hypothesis are required for each inductive step. For example, complete induction can be used to show that

$$F_n = \frac{\varphi^n - \psi^n}{\varphi - \psi}$$

where F_n is the nth Fibonacci number, $\varphi = (1 + \sqrt{5})/2$ (the golden ratio) and $\psi = (1 - \sqrt{5})/2$ are the roots of the polynomial $x^2 - x - 1$. By using the fact that $F_{n+2} = F_{n+1} + F_n$ for each $n \in \mathbf{N}$, the identity above can be verified by direct calculation for F_{n+2} if one assumes that it already holds for both F_{n+1} and F_n. To complete the proof, the identity must be verified in the two base cases $n = 0$ and $n = 1$.

Another proof by complete induction uses the hypothesis that the statement holds for *all* smaller n more thoroughly. Consider the statement that "every natural number greater than 1 is a product of (one or more) prime numbers", and assume that for a given $m > 1$ it holds for all smaller $n > 1$. If m is prime then it is certainly a product of primes, and if not, then by definition it is a product: $m = n_1 n_2$, where neither of the factors is equal to 1; hence neither is equal to m, and so both are smaller than m. The induction hypothesis now applies to n_1 and n_2, so each one is a product of primes. Thus m is a product of products of primes; i.e. a product of primes.

TRANSFINITE INDUCTION

Transfinite induction is an extension of mathematical induction to well-ordered sets, for example to sets of ordinal numbers or cardinal numbers.

Let P(α) be a property defined for all ordinals α. Suppose that whenever P(β) is true for all β < α, then P(α) is also true. Then transfinite induction tells us that P is true for all ordinals.

Usually the proof is broken down into three cases:

- **Zero case:** Prove that P(0) is true.
- **Successor case:** Prove that for any successor ordinal α+1, P(α+1) follows from P(α) (and, if necessary, P(β) for all β < α).
- **Limit case:** Prove that for any limit ordinal λ, P(λ) follows from [P(β) for all β < λ].

All three cases are identical except for the type of ordinal considered. They do not formally need to be considered separately, but in practice the proofs are typically so different as to require separate presentations. Zero is sometimes considered a limit ordinal and then may sometimes be treated in proofs in the same case as limit ordinals.

Transfinite Recursion

Transfinite recursion is similar to transfinite induction; however, instead of proving that something holds for all ordinal numbers, we construct a sequence of objects, one for each ordinal.

As an example, a basis for a (possibly infinite-dimensional) vector space can be created by choosing a vector v_0 and for each ordinal α choosing a vector that is not in the span of the vectors $\{v_\beta | \beta < \alpha\}$. This process stops when no vector can be chosen.

More formally, we can state the Transfinite Recursion Theorem as follows:

- **Transfinite Recursion Theorem (version 1).** Given a class function *G*: *V* → *V* (where *V* is the class of all sets), there exists a unique transfinite sequence *F*: Ord → *V* (where Ord is the class of all ordinals) such that

F(α) = *G*(*F* ↾ α) for all ordinals α, where ↾ denotes the restriction of *F*'s domain to ordinals < α.

As in the case of induction, we may treat different types of ordinals separately: another formulation of transfinite recursion is the following:

- **Transfinite Recursion Theorem (version 2).** Given a set g_1, and class functions G_2, G_3, there exists a unique function *F*: Ord → *V* such that
- $F(0) = g_1$,
- $F(\alpha + 1) = G_2(F(\alpha))$, for all α ∈ Ord,
- $F(\lambda) = G_3(F \restriction \lambda)$, for all limit λ ≠ 0.

Note that we require the domains of G_2, G_3 to be broad enough to make the above properties meaningful. The uniqueness of the sequence satisfying these properties can be proved using transfinite induction.

More generally, one can define objects by transfinite recursion on any well-founded relation R. (R need not even be a set; it can be a proper class, provided it is a set-like relation; that is, for any x, the collection of all y such that $y \, R \, x$ must be a set.)

Relationship to the Axiom of Choice

Proofs or constructions using induction and recursion often use the axiom of choice to produce a well-ordered relation that can be treated by transfinite induction. However, if the relation in question is already well-ordered, one can often use transfinite induction without invoking the axiom of choice. For example, many results about Borel sets are proved by transfinite induction on the ordinal rank of the set; these ranks are already well-ordered, so the axiom of choice is not needed to well-order them.

The following construction of the Vitali set shows one way that the axiom of choice can be used in a proof by transfinite induction:

First, well-order the real numbers (this is where the axiom of choice enters via the well-ordering theorem), giving a sequence $\langle r_\alpha | \alpha < \beta \rangle$, where β is an ordinal with the cardinality of the continuum. Let v_0 equal r_0. Then let v_1 equal $r_{\alpha 1}$, where α_1 is least such that $r_{\alpha 1} - v_0$ is not a rational number. Continue; at each step use the least real from the r sequence that does not have a rational difference with any element thus far constructed in the v sequence. Continue until all the reals in the r sequence are exhausted. The final v sequence will enumerate the Vitali set.

The above argument uses the axiom of choice in an essential way at the very beginning, in order to well-order the reals. After that step, the axiom of choice is not used again.

Other uses of the axiom of choice are more subtle. For example, a construction by transfinite recursion frequently will not specify a *unique* value for $A_{\alpha+1}$, given the sequence up to α, but will specify only a *condition* that $A_{\alpha+1}$ must satisfy, and argue that there is at least one set satisfying this condition. If it is not possible to define a unique example of such a set at each stage, then it may be necessary to invoke (some form of) the axiom of choice to select one such at each step. For inductions and recursions of countable length, the weaker axiom of dependent choice is sufficient. Because there are models of Zermelo–Fraenkel set theory of interest to set theorists that satisfy the axiom of dependent choice but not the full axiom of choice, the knowledge that a particular proof only requires dependent choice can be useful.

The last two steps can be reformulated as one step:

- Showing that if the statement holds for all $n < m$ then the same statement also holds for $n = m$.

This form of mathematical induction is not only valid for statements about natural numbers, but for statements about elements of any well-founded set, that is, a set with an irreflexive relation < that contains no infinite descending chains.

This form of induction, when applied to ordinals (which form a well-ordered and hence well-founded class), is called *transfinite induction*. It is an important proof technique in set theory, topology and other fields.

Proofs by transfinite induction typically distinguish three cases:

- when m is a minimal element, i.e. there is no element smaller than m
- when m has a direct predecessor, i.e. the set of elements which are smaller than m has a largest element
- when m has no direct predecessor, i.e. m is a so-called limit-ordinal

Strictly speaking, it is not necessary in transfinite induction to prove the basis, because it is a vacuous special case of the proposition that if P is true of all $n < m$, then P is true of m. It is vacuously true precisely because there are no values of $n < m$ that could serve as counterexamples.

Equivalence with the Well-Ordering Principle

The principle of mathematical induction is usually stated as an axiom of the natural numbers; see Peano axioms. However, it can be proved from the well-ordering principle. Indeed, suppose the following:

- The set of natural numbers is well-ordered.
- Every natural number is either zero, or $n + 1$ for some natural number n.
- For any natural number n, $n + 1$ is greater than n.

To derive simple induction from these axioms, one must show that if $\mathrm{P}(n)$ is some proposition predicated of n, and if:

- $\mathrm{P}(0)$ holds and
- whenever $\mathrm{P}(k)$ is true then $\mathrm{P}(k + 1)$ is also true

then $\mathrm{P}(n)$ holds for all n.

Proof. Let S be the set of all natural numbers for which $\mathrm{P}(n)$ is false. Let us see what happens if one asserts that S is nonempty. Well-ordering tells us that S has a least element, say t. Moreover, since $\mathrm{P}(0)$ is true, t is not 0. Since every natural number is either zero or some $n + 1$, there is some natural number n such that $n + 1 = t$. Now n is less than t, and t is the least element of S. It follows that n is not in S, and so $\mathrm{P}(n)$ is true. This means that $\mathrm{P}(n + 1)$ is true, and so $\mathrm{P}(t)$ is true. This is a contradiction, since t was in S. Therefore, S is empty.

It can also be proved that induction, given the other axioms, implies the well-ordering principle.

Proof. Suppose there exists a non-empty set, S, of naturals with no least element. Let $P(k)$ be the assertion that k is not in S. Then $P(0)$ is true for if it were false then 0 is the least element of S. Furthermore, suppose $P(1)$, $P(2)$,..., $P(k)$ is true. Then if $P(k+1)$ is false $k+1$ is in S, thus it is the minimal element is S, a contradiction. Thus $P(k+1)$ is true. Therefore by the induction axiom S is empty, a contradiction.

Example of Error in the Inductive Step

This example demonstrated a subtle error in the proof of the inductive step.

Joel E. Cohen proposed the following argument, which purports to prove by mathematical induction that all horses are of the same color:

- Basis: In a set of only *one* horse, there is only one color.
- Induction step: Assume as induction hypothesis that within any set of n horses, there is only one color. Now look at any set of $n + 1$ horses. Number them: 1, 2, 3, ..., n, $n + 1$. Consider the sets $\{1, 2, 3, ..., n\}$ and $\{2, 3, 4, ..., n + 1\}$. Each is a set of only n horses, therefore within each there is only one color. But the two sets overlap, so there must be only one color among all $n + 1$ horses.

The basis case $n = 1$ is trivial (as any horse is the same color as itself), and the inductive step is correct in all cases $n > 1$. However, the logic of the inductive step is incorrect for $n = 1$, because the statement that "the two sets overlap" is false (there are only $n + 1 = 2$ horses prior to either removal, and after removal the sets of one horse each do not overlap).

REFERENCES

1. Anderson, Robert B. (1979). Proving Programs Correct. New York: John Wiley & Sons. p. 1. ISBN 0471033952.
2. Suber, Peter. "Mathematical Induction". Earlham College. Retrieved 26 March 2011.
3. Acerbi, Fabio. "Plato: Parmenides 149a7-c3. A Proof by Complete Induction?". Archive for History of Exact Sciences 55 (2000), 57–76.
4. Chris K. Caldwell. "Euclid's Proof of the Infinitude of Primes (c. 300 BC)". utm.edu. Retrieved 2016-02-28.
5. "Euclid's Primes". mathsisgoodforyou.com. Retrieved 2016-02-28.
6. "Proofs of the Infinity of the Prime Numbers". hermetic.ch. Archived from the original on 2015-01-12.
7. Cajori (1918), p. 197: 'The process of reasoning called "Mathematical Induction" has had several independent origins. It has been traced back to the Swiss Jakob (James) Bernoulli, the Frenchman B. Pascal and P. Fermat, and the Italian F. Maurolycus. [...] By reading a little between the lines one can find traces of mathematical induction still earlier, in the writings of the Hindus and the Greeks, as, for instance, in the "cyclic method" of Bhaskara, and in Euclid's proof that the number of primes is infinite.'
8. Rashed, R. (1994), "Mathematical induction: al-Karajī and al-Samawal", The Development of Arabic Mathematics: Between Arithmetic and Algebra, Boston Studies in the Philosophy of Science, 156, Kluwer Academic Publishers, pp. 62–84, ISBN 9780792325659
9. "It is sometimes required to prove a theorem which shall be true whenever a certain quantity n which it involves shall be an integer or whole number and the method of proof is usually of the following kind. 1st. The theorem is proved to be true when n = 1. 2ndly. It is proved that if the theorem is true when n is a given whole number, it will be true if n is the next greater integer.

Hence the theorem is true universally. . .. This species of argument may be termed a continued sorites" (Boole circa 1849 Elementary Treatise on Logic not mathematical pages 40–41 reprinted in Grattan-Guinness, Ivor and Bornet, Gérard (1997), George Boole: Selected Manuscripts on Logic and its Philosophy, Birkhäuser Verlag, Berlin, ISBN 3-7643-5456-9)

10. Peirce, C. S. (1881). "On the Logic of Number". American Journal of Mathematics. 4 (1–4). pp. 85–95. doi:10.2307/2369151. JSTOR 2369151. MR 1507856. Reprinted (CP 3.252-88), (W 4:299-309).
11. Shields (1997)
12. Hermes (1973), VI.3.1
13. Hermes (1973), VI.4.3, presenting a theorem of Thoralf Skolem
14. Buss, Samuel (1986). Bounded Arithmetic. Naples: Bibliopolis.
15. Cohen, Joel E. (1961), "On the nature of mathematical proof", Opus. Reprinted in A Random Walk in Science (R. L. Weber, ed.), Crane, Russak & Co., 1973.

CHAPTER 3

Mathematical Proof

Chapter Contents

- **Introduction**
- **Methods**
- **Inductive Reasoning**

Sometimes people read mathematical proofs and think they are reading a foreign language. This book describes the language used in a mathematical proof and also the different types of proofs used in math. This book is essential to develop rigorous mathematics.

INTRODUCTION

In mathematics, a **proof** is a deductive argument for a mathematical statement. In the argument, other previously established statements, such as theorems, can be used. In principle, a proof can be traced back to self-evident or assumed statements, known as axioms, along with accepted rules of inference. Axioms may be treated as conditions that must be met before the statement applies. Proofs are examples of deductive reasoning and are distinguished from inductive or empirical arguments; a proof must demonstrate that a statement is always true (occasionally by listing *all* possible cases and showing that it holds in each), rather than enumerate many confirmatory cases. An unproved proposition that is believed to be true is known as a conjecture.

Proofs employ logic but usually include some amount of natural language which usually admits some ambiguity. In fact, the vast majority of proofs in written mathematics can be considered as applications of rigorous informal logic. Purely formal proofs, written in symbolic language instead of natural language, are considered in proof theory. The distinction between formal and informal proofs has led to much examination of current and historical mathematical practice, quasi-empiricism in mathematics, and so-called folk mathematics (in both senses of that term). The philosophy of mathematics is concerned with the role of language and logic in proofs, and mathematics as a language.

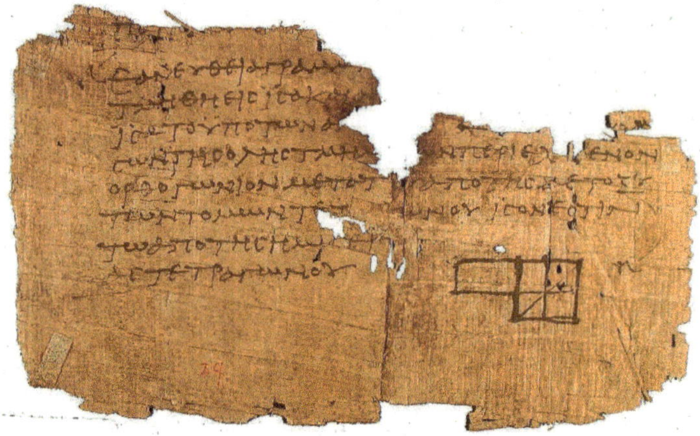

P. Oxy. 29, one of the oldest surviving fragments of Euclid's *Elements*, a textbook used for millennia to teach proof-writing techniques. The diagram accompanies Book II, Proposition 5.

The word "proof" comes from the Latin *probare* meaning "to test". Related modern words are the English "probe", "probation", and "probability", the Spanish *probar* (to smell or taste, or (lesser use) touch or test), Italian *provare* (to try), and the German *probieren* (to try). The early use of "probity" was in the presentation of legal evidence. A person of authority, such as a nobleman, was said to have probity, whereby the evidence was by his relative authority, which outweighed empirical testimony.

Plausibility arguments using heuristic devices such as pictures and analogies preceded strict mathematical proof. It is likely that the idea of demonstrating a conclusion first arose in connection with geometry, which originally meant the same as "land measurement". The development of mathematical proof is primarily the product of ancient Greek mathematics, and one of the greatest achievements thereof. Thales (624–546 BCE) proved some theorems in geometry. Eudoxus (408–355 BCE) and Theaetetus (417–369 BCE) formulated theorems but did not prove them. Aristotle (384–322 BCE) said definitions should describe the concept being defined in terms of other concepts already known. Mathematical proofs were revolutionized by Euclid (300 BCE), who introduced the axiomatic method still in use today, starting with undefined terms and axioms (propositions regarding the undefined terms assumed to be self-evidently true from the Greek "axios" meaning "something worthy"), and used these to prove theorems using deductive logic. His book, the *Elements*, was read by anyone who was considered educated in the West until the middle of the 20th century. In addition to theorems of geometry, such as the

Pythagorean theorem, the *Elements* also covers number theory, including a proof that the square root of two is irrational and that there are infinitely many prime numbers.

Further advances took place in medieval Islamic mathematics. While earlier Greek proofs were largely geometric demonstrations, the development of arithmetic and algebra by Islamic mathematicians allowed more general proofs that no longer depended on geometry. In the 10th century CE, the Iraqi mathematician Al-Hashimi provided general proofs for numbers (rather than geometric demonstrations) as he considered multiplication, division, etc. for "lines." He used this method to provide a proof of the existence of irrational numbers. An inductive proof for arithmetic sequences was introduced in the *Al-Fakhri* (1000) by Al-Karaji, who used it to prove the binomial theorem and properties of Pascal's triangle. Alhazen also developed the method of proof by contradiction, as the first attempt at proving the Euclidean parallel postulate.

Modern proof theory treats proofs as inductively defined data structures. There is no longer an assumption that axioms are "true" in any sense; this allows for parallel mathematical theories built on alternate sets of axioms (see Axiomatic set theory and Non-Euclidean geometry for examples).

Nature and Purpose

As practiced, a proof is expressed in natural language and is a rigorous argument intended to convince the audience of the truth of a statement. The standard of rigor is not absolute and has varied throughout history. A proof can be presented differently depending on the intended audience. In order to gain acceptance, a proof has to meet communal statements of rigor; an argument considered vague or incomplete may be rejected.

The concept of a proof is formalized in the field of mathematical logic. A formal proof is written in a formal language instead of a natural language. A formal proof is defined as sequence of formulas in a formal language, in which each formula is a logical consequence of preceding formulas. Having a definition of formal proof makes the concept of proof amenable to study. Indeed, the field of proof theory studies formal proofs and their properties, for example, the property that a statement has a formal proof. An application of proof theory is to show that certain undecidable statements are not provable.

The definition of a formal proof is intended to capture the concept of proofs as written in the practice of mathematics. The soundness of this definition amounts to the belief that a published proof can, in principle, be converted into a formal proof. However, outside the field of automated proof assistants, this is rarely done in practice. A classic question in philosophy asks whether mathematical proofs are analytic or synthetic. Kant, who introduced the analytic-synthetic distinction, believed mathematical proofs are synthetic.

Proofs may be viewed as aesthetic objects, admired for their mathematical beauty. The mathematician Paul Erdős was known for describing proofs he found particularly elegant as coming from "The Book", a hypothetical tome containing the most beautiful method(s) of proving each theorem. The book *Proofs from THE BOOK*, published in 2003, is devoted to presenting 32 proofs its editors find particularly pleasing.

METHODS

Direct Proof

In direct proof, the conclusion is established by logically combining the axioms, definitions, and earlier theorems. For example, direct proof can be used to establish that the sum of two even integers is always even:

Consider two even integers x and y. Since they are even, they can be written as $x = 2a$ and $y = 2b$, respectively, for integers a and b. Then the sum $x + y = 2a + 2b = 2(a+b)$. Therefore $x+y$ has 2 as a factor and, by definition, is even. Hence the sum of any two even integers is even.

This proof uses the definition of even integers, the integer properties of closure under addition and multiplication, and distributivity.

Proof by Mathematical Induction

Despite its name, mathematical induction is a method of deduction, not a form of inductive reasoning. In proof by mathematical induction, a single "base case" is proved, and an "induction rule" is proved that establishes that any arbitrary case implies the next case. Since in principle the induction rule can be applied repeatedly starting from the proved base case, we see that all (usually infinitely many) cases are provable. This avoids having to prove each case individually. A variant of mathematical induction is proof by infinite descent, which can be used, for example, to prove the irrationality of the square root of two.

A common application of proof by mathematical induction is to prove that a property known to hold for one number holds for all natural numbers:Let $\mathbf{N} = \{1,2,3,4,...\}$ be the set of natural numbers, and $P(n)$ be a mathematical statement involving the natural number n belonging to $\mathbf{N}$ such that

- $P(1)$ is true, i.e., $P(n)$ is true for $n = 1$.
- $P(n+1)$ is true whenever $P(n)$ is true, i.e., $P(n)$ is true implies that $P(n+1)$ is true.
- **Then $P(n)$ is true for all natural numbers n.**

For example, we can prove by induction that all positive integers of the form $2n - 1$ are odd. Let $P(n)$ represent "$2n - 1$ is odd":

(i) For $n = 1$, $2n - 1 = 2(1) - 1 = 1$, and 1 is odd, since it leaves a remainder of 1 when divided by 2. Thus $P(1)$ is true.

(ii) For any n, if $2n - 1$ is odd ($P(n)$), then $(2n - 1) + 2$ must also be odd, because adding 2 to an odd number results in an odd number. But $(2n - 1) + 2 = 2n + 1 = 2(n+1) - 1$, so $2(n+1) - 1$ is odd ($P(n+1)$). So $P(n)$ implies $P(n+1)$.

Thus $2n - 1$ is odd, for all positive integers n.

The shorter phrase "proof by induction" is often used instead of "proof by mathematical induction".

Proof by Contraposition

Proof by contraposition infers the conclusion "if p then q" from the premise "if *not* q then *not* p". The statement "if *not* q then *not* p" is called the contrapositive of the statement "if p then q". For example, contraposition can be used to establish that, given an integer x, if x^2 is even, then x is even:

Suppose x is not even. Then x is odd. The product of two odd numbers is odd, hence $x^2 = x \cdot x$ is odd. Thus x^2 is not even. Thus, if x^2 *is* even, the supposition must be false, so x has to be even.

Proof by Contradiction

In proof by contradiction (also known as *reductio ad absurdum*, Latin for "by reduction to the absurd"), it is shown that if some statement were true, a logical contradiction occurs, hence the statement must be false. A famous example of proof by contradiction shows that is an irrational number:

Suppose that $\sqrt{2}$ were a rational number, so by definition $\sqrt{2} = \frac{a}{b}$ where a and b are non-zero integers with no common factor. (If there is a common factor, divide both numerator and denominator by that factor to remove it, and repeat until no common factor remains. By the method of infinite descent, this process must terminate.) Thus, $b\sqrt{2} = a$ Squaring both sides yields $2b^2 = a^2$. Since 2 divides the left hand side, 2 must also divide the right hand side (otherwise an even number would equal an odd number). So a^2 is even, which implies that a must also be even. So we can write $a = 2c$, where c is also an integer. Substitution into the original equation yields $2b^2 = (2c)^2 = 4c^2$. Dividing both sides by 2 yields $b^2 = 2c^2$. But then, by the same argument as before, 2 divides b^2, so b must be even. However, if a and b are both even, they have a common factor, namely 2. This contradicts our initial supposition, so we are forced to conclude that $\sqrt{2}$ is an irrational number.

Proof by Construction

Proof by construction, or proof by example, is the construction of a concrete example with a property to show that something having that property exists. Joseph Liouville, for instance, proved the existence of transcendental numbers by constructing an explicit example. It can also be used to construct a counterexample to disprove a proposition that all elements have a certain property.

Proof by Exhaustion

In proof by exhaustion, the conclusion is established by dividing it into a finite number of cases and proving each one separately. The number of cases sometimes can become very large. For example, the first proof of the four color theorem was a proof by exhaustion with 1,936 cases. This proof was controversial because the majority of the cases were checked by a computer program, not by hand. The shortest known proof of the four color theorem as of 2011 still has over 600 cases.

Probabilistic Proof

A probabilistic proof is one in which an example is shown to exist, with certainty, by using methods of probability theory. Probabilistic proof, like proof by construction, is one of many ways to show existence theorems.

This is not to be confused with an argument that a theorem is 'probably' true, a 'plausibility argument'. The work on the Collatz conjecture shows how far plausibility is from genuine proof.

Combinatorial Proof

A combinatorial proof establishes the equivalence of different expressions by showing that they count the same object in different ways. Often a bijection between two sets is used to show that the expressions for their two sizes are equal. Alternatively, a double counting argument provides two different expressions for the size of a single set, again showing that the two expressions are equal.

Nonconstructive Proof

A nonconstructive proof establishes that a mathematical object with a certain property exists without explaining how such an object can be found. Often, this takes the form of a proof by contradiction in which the nonexistence of the object is proved to be impossible. In contrast, a constructive proof establishes that a particular object exists by providing a method of finding it. A famous example of a nonconstructive proof shows that there exist two irrational numbers a and b such that a^b is a rational number:

Either $\sqrt{2}^{\sqrt{2}}$ is a rational number and we are done (take $a = b = \sqrt{2}$), or $\sqrt{2}^{\sqrt{2}}$ is irrational so we can write $a = \sqrt{2}^{\sqrt{2}}$ and $b = \sqrt{2}$. This then gives $\left(\sqrt{2}^{\sqrt{2}}\right)^{\sqrt{2}} = \sqrt{2}^2 = 2$, which is thus a rational of the form a^b.

Statistical Proofs in Pure Mathematics

The expression "statistical proof" may be used technically or colloquially in areas of pure mathematics, such as involving cryptography, chaotic series, and probabilistic or analytic number theory. It is less commonly used to refer to a mathematical proof in the branch of mathematics known as mathematical statistics. See also "Statistical proof using data" section below.

Computer-Assisted Proofs

Until the twentieth century it was assumed that any proof could, in principle, be checked by a competent mathematician to confirm its validity.However, computers are now used both to prove theorems and to carry out calculations that are too long for any human or team of humans to check; the first proof of the four color theorem is an example of a computer-assisted proof. Some mathematicians are concerned that the possibility of an error in a computer program or a run-time error in its calculations calls the validity of such computer-assisted proofs into question. In practice, the chances of an error invalidating a computer-assisted proof can be reduced by incorporating redundancy and self-checks into calculations, and by developing multiple independent approaches and programs. Errors can never be completely ruled out in case of verification of a proof by humans either, especially if the proof contains natural language and requires deep mathematical insight.

Undecidable Statements

A statement that is neither provable nor disprovable from a set of axioms is called undecidable (from those axioms). One example is the parallel postulate, which is neither provable nor refutable from the remaining axioms of Euclidean geometry.

Mathematicians have shown there are many statements that are neither provable nor disprovable in Zermelo-Fraenkel set theory with the axiom of choice (ZFC), the standard system of set theory in mathematics (assuming that ZFC is consistent); see list of statements undecidable in ZFC.

Gödel's (first) incompleteness theorem shows that many axiom systems of mathematical interest will have undecidable statements.

Heuristic Mathematics and Experimental Mathematics

While early mathematicians such as Eudoxus of Cnidus did not use proofs, from Euclid to the foundational mathematics developments of the late 19th and 20th centuries, proofs were an essential part of mathematics. With the increase in computing power in the 1960s, significant work began to be done investigating mathematical objects outside of the proof-theorem framework, in experimental mathematics. Early pioneers of these methods intended the work ultimately to be embedded in a classical proof-theorem framework, e.g. the early development of fractal geometry, which was ultimately so embedded.

Related Concepts

Visual Proof

Although not a formal proof, a visual demonstration of a mathematical theorem is sometimes called a "proof without words". The left-hand picture below is an example of a historic visual proof of the Pythagorean theorem in the case of the (3,4,5) triangle.

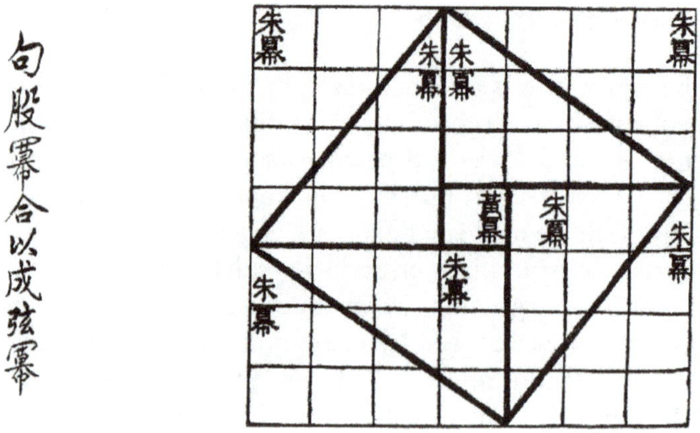

Visual proof for the (3, 4, 5) triangle as in the Zhoubi Suanjing 500–200 BC.

Animated visual proof for the Pythagorean theorem by rearrangement.

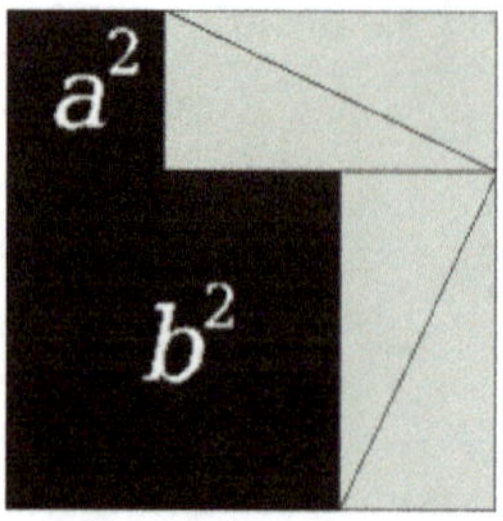

A second animated proof of the Pythagorean theorem.

Some illusory visual proofs, such as the missing square puzzle, can be constructed in a way which appear to prove a supposed mathematical fact but only do so under the presence of tiny errors (for example, supposedly straight lines which actually bend slightly) which are unnoticeable until the entire picture is closely examined, with lengths and angles precisely measured or calculated.

Elementary Proof

An elementary proof is a proof which only uses basic techniques. More specifically, the term is used in number theory to refer to proofs that make no use of complex analysis. For some time it was thought that certain theorems, like the prime number theorem, could only be proved using "higher" mathematics. However, over time, many of these results have been reproved using only elementary techniques.

Two-Column Proof

Proposition XIX. Theorem

If two paralled lines are cut by a transversal, the corresponding angles are equal.
[Converse of Prop. XIV.]

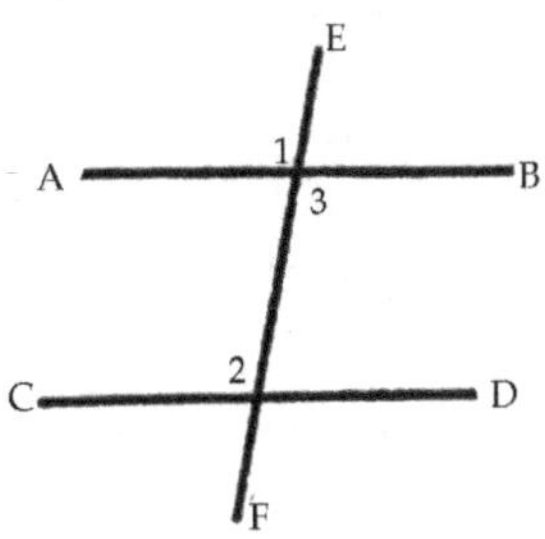

Given parallel lines AB and CD and the cor. Δ 1 and 2.
Tp prove ∠1 = ∠2.

Proof

Statements	Reasons
∠1 = ∠3.	Vertical D are equal.
∠2 = ∠3.	Alt. int. D of lines are equal.
∴ ∠1 = ∠2.	Things equal to the same thing are equal to each other.

Q.R.D.

A two-column proof published in 1913

A particular way of organising a proof using two parallel columns is often used in elementary geometry classes in the United States. The proof is written as a series of lines in two columns. In each line, the left-hand column contains a proposition, while the right-hand column contains a brief explanation of how the corresponding proposition in the left-hand column is either an axiom, a hypothesis, or can be logically derived from previous propositions. The left-hand column is typically headed "Statements" and the right-hand column is typically headed "Reasons".

Colloquial Use of "Mathematical Proof"

The expression "mathematical proof" is used by lay people to refer to using mathematical methods or arguing with mathematical objects, such as numbers, to demonstrate something about everyday life, or when data used in an argument is numerical. It is sometimes also used to mean a "statistical proof" (below), especially when used to argue from data.

Statistical Proof Using Data

"Statistical proof" from data refers to the application of statistics, data analysis, or Bayesian analysis to infer propositions regarding the probability of data. While *using* mathematical proof to establish theorems in statistics, it is usually not a mathematical proof in that the *assumptions* from which probability statements are derived require empirical evidence from outside mathematics to verify. In physics, in addition to statistical methods, "statistical proof" can refer to the specialized *mathematical methods of physics* applied to analyze data in a particle physics experiment or observational study in physical cosmology. "Statistical proof" may also refer to raw data or a convincing diagram involving data, such as scatter plots, when the data or diagram is adequately convincing without further analysis.

Inductive Logic Proofs and Bayesian Analysis

Proofs using inductive logic, while considered mathematical in nature, seek to establish propositions with a degree of certainty, which acts in a similar manner to probability, and may be less than full certainty. Inductive logic should not be confused with mathematical induction.

Bayesian analysis uses Bayes' theorem to update a person's assessment of likelihoods of hypotheses when new evidence or information is acquired.

Proofs as Mental Objects

Psychologism views mathematical proofs as psychological or mental objects. Mathematician philosophers, such as Leibniz, Frege, and Carnap have variously criticized this view and attempted to develop a semantics for what they considered to be the language of thought, whereby standards of mathematical proof might be applied to empirical science.

Influence of Mathematical Proof Methods Outside Mathematics

Philosopher-mathematicians such as Spinoza have attempted to formulate philosophical arguments in an axiomatic manner, whereby mathematical proof standards could be applied to argumentation in general philosophy. Other mathematician-philosophers have tried to use standards of mathematical proof and reason, without empiricism, to arrive at statements outside of mathematics, but having the certainty of propositions deduced in a mathematical proof, such as Descartes' *cogito* argument.

Ending a Proof

Sometimes, the abbreviation *"Q.E.D."* is written to indicate the end of a proof. This abbreviation stands for *«Quod Erat Demonstrandum"*, which is Latin for *"that which was to be demonstrated"*. A more common alternative is to use a square or a rectangle, such as □ or ■, known as a "tombstone" or "halmos" after its eponym Paul Halmos. Often, "which was to be shown" is verbally stated when writing "QED", "□", or "■" during an oral presentation.

INDUCTIVE REASONING

Inductive reasoning (as opposed to *deductive* reasoning or *abductive* reasoning) is reasoning in which the premises are viewed as supplying strong evidence for the truth of the conclusion. While the conclusion of a deductive argument is certain, the truth of the conclusion of an inductive argument is *probable*, based upon the evidence given.

Many dictionaries define inductive reasoning as the derivation of general principles from specific observations, though some sources disagree with this usage.

The philosophical definition of inductive reasoning is more nuanced than simple progression from particular/individual instances to broader generalizations. Rather, the premises of an inductive logical argument indicate some degree of support (inductive probability) for the conclusion but do not entail it; that is, they suggest truth but do not ensure it. In this manner, there is the possibility of moving from general statements to individual instances (for example, statistical syllogisms, discussed below).

Description

Inductive reasoning is inherently uncertain. It only deals in degrees to which, given the premises, the conclusion is *credible* according to some theory of evidence. Examples include a many-valued logic, Dempster–Shafer theory, or probability theory with rules for inference such as Bayes' rule. Unlike deductive reasoning, it does not rely on universals holding over a closed domain of discourse to draw conclusions, so it can be applicable even in cases of epistemic uncertainty (technical issues with this may arise however; for example, the second axiom of probability is a closed-world assumption).

An example of an inductive argument:

All biological life forms that we know of depend on liquid water to exist.

Therefore, if we discover a new biological life form it will probably depend on liquid water to exist.

This argument could have been made every time a new biological life form was found, and would have been correct every time; however, it is still possible that in the future a biological life form not requiring liquid water could be discovered.

As a result, the argument may be stated less formally as:

All biological life forms that we know of depend on liquid water to exist.

All biological life probably depends on liquid water to exist.

Inductive Vs. Deductive Reasoning

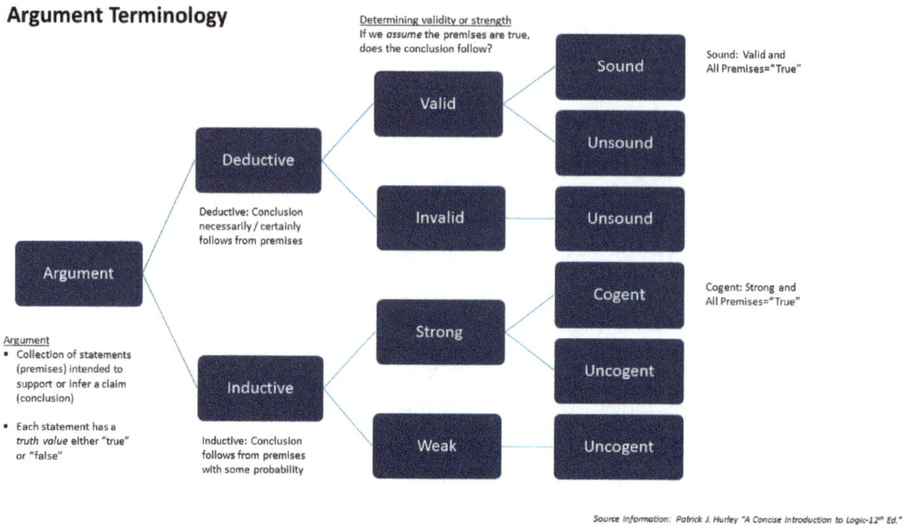

Argument terminology

Unlike deductive arguments, inductive reasoning allows for the possibility that the conclusion is false, even if all of the premises are true. Instead of being valid or invalid, inductive arguments are either *strong* or *weak*, which describes how *probable* it is that the conclusion is true. Another crucial difference is that deductive certainty is impossible in non-axiomatic systems, such as reality, leaving inductive reasoning as the primary route to (probabilistic) knowledge of such systems.

Given that "if A is true then that would cause B, C, and D to be true", an example of deduction would be "A is true therefore we can deduce that B, C, and D are true". An example of induction would be "B, C, and D are observed to be true therefore A might be true". A is a reasonable explanation for B, C, and D being true.

For example:

A large enough asteroid impact would create a very large crater and cause a severe impact winter that could drive the non-avian dinosaurs to extinction.

We observe that there is a very large crater in the Gulf of Mexico dating to very near the time of the extinction of the non-avian dinosaurs

Therefore it is possible that this impact could explain why the non-avian dinosaurs became extinct.

Note however that this is not necessarily the case. Other events also coincide with the extinction of the non-avian dinosaurs. For example, the Deccan Traps in India.

A classical example of an *incorrect* inductive argument was presented by John Vickers:

All of the swans we have seen are white.

Therefore, all swans are white. (Or more precisely, "We expect that all swans are white")

The definition of *inductive* reasoning described in this article excludes mathematical induction, which is a form of *deductive* reasoning that is used to strictly prove properties of recursively defined sets.

Criticism

Inductive reasoning has been criticized by thinkers as diverse as Sextus Empiricus and Karl Popper.

The classic philosophical treatment of the problem of induction was given by the Scottish philosopher David Hume.

Although the use of inductive reasoning demonstrates considerable success, its application has been questionable. Recognizing this, Hume highlighted the fact that our mind draws uncertain conclusions from relatively limited experiences. In deduction, the truth value of the conclusion is based on the truth of the premise. In induction, however, the dependence on the premise is always uncertain. As an example, let's assume "all ravens are black." The fact that there are numerous black ravens supports the assumption. However, the assumption becomes inconsistent with the fact that there are white ravens. Therefore, the general rule of "all ravens are black" is inconsistent with the existence of the white raven. Hume further argued that it is impossible to justify inductive reasoning: specifically, that it cannot be justified deductively, so our only option is to justify it inductively. Since this is circular he concluded that our use of induction is unjustifiable with the help of Hume's Fork.

However, Hume then stated that even if induction were proved unreliable, we would still have to rely on it. So instead of a position of severe skepticism, Hume advocated a practical skepticism based on common sense, where the inevitability of induction is accepted.

Bertrand Russell illustrated his skepticism in a story about a turkey, fed every morning without fail, who following the laws of induction concludes this will continue, but then his throat is cut on Thanksgiving day.

Biases

Inductive reasoning is also known as hypothesis construction because any conclusions made are based on current knowledge and predictions. As with deductive arguments, biases can distort the proper application of inductive argument, thereby preventing the reasoner from forming the most logical conclusion based on the clues. Examples of these biases include the availability heuristic, confirmation bias, and the predictable-world bias.

The availability heuristic causes the reasoner to depend primarily upon information that is readily available to him/her. People have a tendency to rely on information that is easily accessible in the world around them. For example, in surveys, when people are asked to estimate the percentage of people who died from various causes, most respondents would choose the causes that have been most prevalent in the media such as terrorism, and murders, and airplane accidents rather than causes such as disease and traffic accidents, which have been technically "less accessible" to the individual since they are not emphasized as heavily in the world around him/her.

The confirmation bias is based on the natural tendency to confirm rather than to deny a current hypothesis. Research has demonstrated that people are inclined to seek solutions to problems that are more consistent with known hypotheses rather than attempt to refute those hypotheses. Often, in experiments, subjects will ask questions that seek answers that fit established hypotheses, thus confirming these hypotheses. For example, if it is hypothesized that Sally is a sociable individual, subjects will naturally seek to confirm the premise by asking questions that would produce answers confirming that Sally is in fact a sociable individual.

The predictable-world bias revolves around the inclination to perceive order where it has not been proved to exist, either at all or at a particular level of abstraction. Gambling, for example, is one of the most popular examples of predictable-world bias. Gamblers often begin to think that they see simple and obvious patterns in the outcomes and, therefore, believe that they are able to predict outcomes based upon what they have witnessed. In reality, however, the outcomes of these games are difficult to predict and highly complex in nature. However, in general, people tend to seek some type of simplistic order to explain or justify their beliefs and experiences, and it is often difficult for them to realise that their perceptions of order may be entirely different from the truth.

Types

Generalization

A generalization (more accurately, an *inductive generalization*) proceeds from a premise about a sample to a conclusion about the population.

The proportion Q of the sample has attribute A.

Therefore:

The proportion Q of the population has attribute A.

Example

There are 20 balls—either black or white—in an urn. To estimate their respective numbers, you draw a sample of four balls and find that three are black and one is white. A good inductive generalization would be that there are 15 black and five white balls in the urn.

How much the premises support the conclusion depends upon (a) the number in the sample group, (b) the number in the population, and (c) the degree to which the sample represents the population (which may be achieved by taking a random sample). The hasty generalization and the biased sample are generalization fallacies.

Statistical Syllogism

A statistical syllogism proceeds from a generalization to a conclusion about an individual.

A proportion Q of population P has attribute A.

An individual X is a member of P.

Therefore:

There is a probability which corresponds to Q that X has A.

The proportion in the first premise would be something like "3/5ths of", "all", "few", etc. Two dicto simpliciter fallacies can occur in statistical syllogisms: "accident" and "converse accident".

Simple Induction

Simple induction proceeds from a premise about a sample group to a conclusion about another individual.

Proportion Q of the known instances of population P has attribute A.

Individual I is another member of P.

Therefore:

There is a probability corresponding to Q that I has A.

This is a combination of a generalization and a statistical syllogism, where the conclusion of the generalization is also the first premise of the statistical syllogism.

Argument from Analogy

The process of analogical inference involves noting the shared properties of two or more things, and from this basis inferring that they also share some further property:

P and Q are similar in respect to properties a, b, and c.

Object P has been observed to have further property x.

Therefore, Q probably has property x also.

Analogical reasoning is very frequent in common sense, science, philosophy and the humanities, but sometimes it is accepted only as an auxiliary method. A refined approach is case-based reasoning.

Causal Inference

A causal inference draws a conclusion about a causal connection based on the conditions of the occurrence of an effect. Premises about the correlation of two things can indicate a causal relationship between them, but additional factors must be confirmed to establish the exact form of the causal relationship.

Prediction

A prediction draws a conclusion about a future individual from a past sample.

Proportion Q of observed members of group G have had attribute A.

Therefore:

There is a probability corresponding to Q that other members of group G will have attribute A when next observed.

Bayesian Inference

As a logic of induction rather than a theory of belief, Bayesian inference does not determine which beliefs are *a priori* rational, but rather determines how we should rationally change the beliefs we have when presented with evidence. We begin by committing to a prior probability for a hypothesis based on logic or previous experience, and when faced with evidence, we adjust the strength of our belief in that hypothesis in a precise manner using Bayesian logic.

Inductive Inference

Around 1960, Ray Solomonoff founded the theory of universal inductive inference, the theory of prediction based on observations; for example, predicting the next symbol based upon a given series of symbols. This is a formal inductive framework that combines algorithmic information theory with the Bayesian framework. Universal inductive inference is based on solid philosophical foundations, and can be considered as a mathematically formalized Occam's razor. Fundamental ingredients of the theory are the concepts of algorithmic probability and Kolmogorov complexity.

REFERENCES

1. Bill Casselman. "One of the Oldest Extant Diagrams from Euclid". University of British Columbia. Retrieved 2008-09-26.
2. Clapham, C. & Nicholson, JN. The Concise Oxford Dictionary of Mathematics, Fourth edition. A statement whose truth is either to be taken as self-evident or to be assumed. Certain areas of mathematics involve choosing a set of axioms and discovering what results can be derived from them, providing proofs for the theorems that are obtained.
3. Cupillari, Antonella. The Nuts and Bolts of Proofs. Academic Press, 2001. Page 3.
4. Gossett, Eric. Discrete Mathematics with Proof. John Wiley and Sons, 2009. Definition 3.1 page 86. ISBN 0-470-45793-7
5. New Shorter Oxford English Dictionary, 1993, OUP, Oxford.
6. The Emergence of Probability, Ian Hacking
7. The History and Concept of Mathematical Proof, Steven G. Krantz. 1. February 5, 2007

8. Kneale, p. 2
9. Howard Eves, An Introduction to the History of Mathematics, Saunders, 1990, ISBN 0-03-029558-0 p. 141: "No work, except The Bible, has been more widely used...."
10. Matvievskaya, Galina (1987), "The Theory of Quadratic Irrationals in Medieval Oriental Mathematics", Annals of the New York Academy of Sciences, 500: 253–277 , doi:10.1111/j.1749-6632.1987.tb37206.x
11. Eder, Michelle (2000), Views of Euclid's Parallel Postulate in Ancient Greece and in Medieval Islam, Rutgers University, retrieved 2008-01-23
12. Buss, Samuel R. (1998), "An introduction to proof theory", in Buss, Samuel R., Handbook of Proof Theory, Studies in Logic and the Foundations of Mathematics, 137, Elsevier, pp. 1–78, ISBN 9780080533186. See in particular p. 3: "The study of Proof Theory is traditionally motivated by the problem of formalizing mathematical proofs; the original formulation of first-order logic by Frege was the first successful step in this direction."
13. Cupillari, page 20.
14. Cupillari, page 46.
15. Examples of simple proofs by mathematical induction for all natural numbers
16. Proof by induction, University of Warwick Glossary of Mathematical Terminology
17. While most mathematicians do not think that probabilistic evidence ever counts as a genuine mathematical proof, a few mathematicians and philosophers have argued that at least some types of probabilistic evidence (such as Rabin's probabilistic algorithm for testing primality) are as good as genuine mathematical proofs. See, for example, Davis, Philip J. (1972), "Fidelity in Mathematical Discourse: Is One and One Really Two?" American Mathematical Monthly 79:252-63. Fallis, Don (1997), "The Epistemic Status of Probabilistic Proof." Journal of Philosophy 94:165-86.
18. "in number theory and commutative algebra... in particular the statistical proof of the lemma."
19. "Whether constant π (i.e., pi) is normal is a confusing problem without any strict theoretical demonstration except for some statistical proof"" (Derogatory use.)
20. "these observations suggest a statistical proof of Goldbach's conjecture with very quickly vanishing probability of failure for large E"
21. "What to do with the pictures? Two thoughts surfaced: the first was that they were unpublishable in the standard way, there were no theorems only very suggestive pictures. They furnished convincing evidence for many conjectures and lures to further exploration, but theorems were coins of the realm ant the conventions of that day dictated that journals only published theorems", David Mumford, Caroline Series and David Wright, Indra's Pearls, 2002
22. "Mandelbrot, working at the IBM Research Laboratory, did some computer simulations for these sets on the reasonable assumption that, if you wanted

to prove something, it might be helpful to know the answer ahead of time."A Note on the History of Fractals,

23. "... brought home again to Benoit Mandelbrot that there was a 'mathematics of the eye', that visualization of a problem was as valid a method as any for finding a solution. Amazingly, he found himself alone with this conjecture. The teaching of mathematics in France was dominated by a handful of dogmatic mathematicians hiding behind the pseudonym 'Bourbaki'... ", Introducing Fractal Geometry, Nigel Lesmoir-Gordon
24. Patricio G. Herbst, Establishing a Custom of Proving in American School Geometry: Evolution of the Two-Column Proof in the Early Twentieth Century, Educational Studies in Mathematics, Vol. 49, No. 3 (2002), pp. 283-312,
25. Introduction to the Two-Column Proof, Carol Fisher
26. Copi, I. M.; Cohen, C.; Flage, D. E. (2007). Essentials of Logic (Second ed.). Upper Saddle River, NJ: Pearson Education. ISBN 978-0-13-238034-8.
27. "Deductive and Inductive Arguments", Internet Encyclopedia of Philosophy, Some dictionaries define "deduction" as reasoning from the general to specific and "induction" as reasoning from the specific to the general. While this usage is still sometimes found even in philosophical and mathematical contexts, for the most part, it is outdated.
28. Kosko, Bart (1990). "Fuzziness vs. Probability". International Journal of General Systems. 17 (1): 211–240. doi:10.1080/03081079008935108.
29. John Vickers. The Problem of Induction. The Stanford Encyclopedia of Philosophy.
30. Herms, D. "Logical Basis of Hypothesis Testing in Scientific Research" (pdf).
31. "Stanford Encyclopedia of Philosophy : Kant's account of reason".
32. Chowdhry, K.R. (January 2, 2015). Fundamentals of Discrete Mathematical Structures (3rd ed.). PHI Learning Pvt. Ltd. p. 26. Retrieved 1 December 2016.
33. Sextus Empiricus, Outlines Of Pyrrhonism. Trans. R.G. Bury, Harvard University Press, Cambridge, Massachusetts, 1933, p. 283.
34. Popper, Karl R.; Miller, David W. (1983). "A proof of the impossibility of inductive probability". Nature. 302 (5910): 687–688. doi:10.1038/302687a0.
35. David Hume (1910) . An Enquiry concerning Human Understanding. P.F. Collier & Son. ISBN 0-19-825060-6.
36. Vickers, John. "The Problem of Induction" (Section 2). Stanford Encyclopedia of Philosophy. 21 June 2010
37. Vickers, John. "The Problem of Induction" (Section 2.1). Stanford Encyclopedia of Philosophy. 21 June 2010.
38. The story by Russell is found in Alan Chalmers, What is this thing Called Science, Open University Press, Milton Keynes, 1982, p. 14
39. Gray, Peter (2011). Psychology (Sixth ed.). New York: Worth. ISBN 978-1-4292-1947-1.

40. Baronett, Stan (2008). Logic. Upper Saddle River, NJ: Pearson Prentice Hall. pp. 321–325.
41. For more information on inferences by analogy, see Juthe, 2005.
42. Rathmanner, Samuel; Hutter, Marcus (2011). "A Philosophical Treatise of Universal Induction". Entropy. 13 (6): 1076–1136. doi:10.3390/e13061076.

Axioms

Chapter Contents

Axiom (noun), axiomatic (adjective): via Latin, from Greek axioma, "that which is thought fitting; decision; self-evident principle." The Indo-European root is ag- "to drive, to lead." A subsidiary Greek meaning, "to weigh," led to axioma, literally "something weighty." In mathematical terms, axioms are concepts felt weighty enough that you can base a logical system on them.

INTRODUCTION

An axiom or postulate is a statement that is taken to be true, to serve as a premise or starting point for further reasoning and arguments. The word comes from the Greek *axíōma* (ξίωμα) 'that which is thought worthy or fit' or 'that which commends itself as evident.'

The term has subtle differences in definition when used in the context of different fields of study. As defined in classic philosophy, an axiom is a statement that is so evident or well-established, that it is accepted without controversy or question. As used in modern logic, an axiom is simply a premise or starting point for reasoning.

As used in mathematics, the term *axiom* is used in two related but distinguishable senses: "logical axioms" and "non-logical axioms". Logical axioms are usually statements that are taken to be true within the system of logic they define (e.g., (A and B) implies A), often shown in symbolic form, while non-logical axioms (e.g., $a + b = b + a$) are actually substantive assertions about the elements of the domain of a specific mathematical theory (such as arithmetic). When used in the latter sense, "axiom", "postulate", and "assumption" may be used interchangeably. In general, a non-logical axiom is not a self-evident truth, but rather a formal logical expression used in deduction to build a mathematical theory. As modern mathematics admits multiple, equally "true" systems of logic, precisely the same thing must be said for logical axioms - they both define and are specific to the particular system of logic that is being invoked. To axiomatize a system of knowledge is to show that its claims can be derived from a small, well-understood set of sentences (the axioms). There are typically multiple ways to axiomatize a given mathematical domain.

In both senses, an axiom is any mathematical statement that serves as a starting point from which other statements are logically derived. Within the system they define, axioms (unless redundant) cannot be derived by principles of deduction, nor are they demonstrable by mathematical proofs, simply because they are starting points; there is nothing else from which they logically follow otherwise they would be classified as theorems. However, an axiom in one system may be a theorem in another, and vice versa. Whether it is meaningful (and, if so, what it means) for an axiom, or any mathematical statement, to be "true" is an open question in the philosophy of mathematics.

ETYMOLOGY

The word *axiom* comes from the Greek word ξίωμα (*axíōma*), a verbal noun from the verb ξιόειν (*axioein*), meaning "to deem worthy", but also "to require", which in turn comes from ξιος (*áxios*), meaning "being in balance", and hence "having (the same) value (as)", "worthy", "proper". Among the ancient Greek philosophers an axiom was a claim which could be seen to be true without any need for proof.

The root meaning of the word *postulate* is to "demand"; for instance, Euclid demands that one agree that some things can be done, e.g. any two points can be joined by a straight line, etc.

Ancient geometers maintained some distinction between axioms and postulates. While commenting on Euclid's books, Proclus remarks that, "Geminus held that this 4th Postulate should not be classed as a postulate but as an axiom, since it does not, like the first three Postulates, assert the possibility of some construction but expresses an essential property." Boethius translated ‹postulate' as *petitio* and called the axioms *notiones communes* but in later manuscripts this usage was not always strictly kept.

Early Greeks

The logico-deductive method whereby conclusions (new knowledge) follow from premises (old knowledge) through the application of sound arguments (syllogisms, rules of inference), was developed by the ancient Greeks, and has become the core principle of modern mathematics. Tautologies excluded, nothing can be deduced if nothing is assumed. Axioms and postulates are the basic assumptions underlying a given body of deductive knowledge. They are accepted without demonstration. All other assertions (theorems, if we are talking about mathematics) must be proven with the aid of these basic assumptions. However, the interpretation of mathematical knowledge has changed from ancient times to the modern, and consequently the terms *axiom* and *postulate* hold a slightly different meaning for the present day mathematician, than they did for Aristotle and Euclid.

The ancient Greeks considered geometry as just one of several sciences, and held the theorems of geometry on par with scientific facts. As such, they developed and used the logico-deductive method as a means of avoiding error, and for structuring and communicating knowledge. Aristotle's posterior analytics is a definitive exposition of the classical view.

An "axiom", in classical terminology, referred to a self-evident assumption common to many branches of science. A good example would be the assertion that

When an equal amount is taken from equals, an equal amount results.

At the foundation of the various sciences lay certain additional hypotheses which were accepted without proof. Such a hypothesis was termed a *postulate*. While the axioms were common to many sciences, the postulates of each particular science were different. Their validity had to be established by means of real-world experience. Indeed, Aristotle warns that the content of a science cannot be successfully communicated, if the learner is in doubt about the truth of the postulates.

The classical approach is well-illustrated by Euclid's Elements, where a list of postulates is given (common-sensical geometric facts drawn from our experience), followed by a list of "common notions" (very basic, self-evident assertions).

Postulates

- It is possible to draw a straight line from any point to any other point.
- It is possible to extend a line segment continuously in both directions.
- It is possible to describe a circle with any center and any radius.
- It is true that all right angles are equal to one another.
- ("Parallel postulate") It is true that, if a straight line falling on two straight lines make the interior angles on the same side less than two right angles, the two straight lines, if produced indefinitely, intersect on that side on which are the angles less than the two right angles.

Common Notions

- Things which are equal to the same thing are also equal to one another.
- If equals are added to equals, the wholes are equal.

- If equals are subtracted from equals, the remainders are equal.
- Things which coincide with one another are equal to one another.
- The whole is greater than the part.

Modern Development

A lesson learned by mathematics in the last 150 years is that it is useful to strip the meaning away from the mathematical assertions (axioms, postulates, propositions, theorems) and definitions. One must concede the need for primitive notions, or undefined terms or concepts, in any study. Such abstraction or formalization makes mathematical knowledge more general, capable of multiple different meanings, and therefore useful in multiple contexts. Alessandro Padoa, Mario Pieri, and Giuseppe Peano were pioneers in this movement.

Structuralist mathematics goes further, and develops theories and axioms (e.g. field theory, group theory, topology, vector spaces) without *any* particular application in mind. The distinction between an "axiom" and a "postulate" disappears. The postulates of Euclid are profitably motivated by saying that they lead to a great wealth of geometric facts. The truth of these complicated facts rests on the acceptance of the basic hypotheses. However, by throwing out Euclid's fifth postulate we get theories that have meaning in wider contexts, hyperbolic geometry for example. We must simply be prepared to use labels like "line" and "parallel" with greater flexibility. The development of hyperbolic geometry taught mathematicians that postulates should be regarded as purely formal statements, and not as facts based on experience.

When mathematicians employ the field axioms, the intentions are even more abstract. The propositions of field theory do not concern any one particular application; the mathematician now works in complete abstraction. There are many examples of fields; field theory gives correct knowledge about them all.

It is not correct to say that the axioms of field theory are "propositions that are regarded as true without proof." Rather, the field axioms are a set of constraints. If any given system of addition and multiplication satisfies these constraints, then one is in a position to instantly know a great deal of extra information about this system.

Modern mathematics formalizes its foundations to such an extent that mathematical theories can be regarded as mathematical objects, and mathematics itself can be regarded as a branch of logic. Frege, Russell, Poincaré, Hilbert, and Gödel are some of the key figures in this development.

In the modern understanding, a set of axioms is any collection of formally stated assertions from which other formally stated assertions follow by the application of certain well-defined rules. In this view, logic becomes just another formal system. A set of axioms should be consistent; it should be impossible to derive a contradiction from the axiom. A set of axioms should also be non-redundant; an assertion that can be deduced from other axioms need not be regarded as an axiom.

It was the early hope of modern logicians that various branches of mathematics, perhaps all of mathematics, could be derived from a consistent collection of basic axioms.

An early success of the formalist program was Hilbert's formalization of Euclidean geometry, and the related demonstration of the consistency of those axioms.

In a wider context, there was an attempt to base all of mathematics on Cantor's set theory. Here the emergence of Russell's paradox, and similar antinomies of naïve set theory raised the possibility that any such system could turn out to be inconsistent.

The formalist project suffered a decisive setback, when in 1931 Gödel showed that it is possible, for any sufficiently large set of axioms (Peano's axioms, for example) to construct a statement whose truth is independent of that set of axioms. As a corollary, Gödel proved that the consistency of a theory like Peano arithmetic is an unprovable assertion within the scope of that theory.

It is reasonable to believe in the consistency of Peano arithmetic because it is satisfied by the system of natural numbers, an infinite but intuitively accessible formal system. However, at present, there is no known way of demonstrating the consistency of the modern Zermelo–Fraenkel axioms for set theory. Furthermore, using techniques of forcing (Cohen) one can show that the continuum hypothesis (Cantor) is independent of the Zermelo–Fraenkel axioms. Thus, even this very general set of axioms cannot be regarded as the definitive foundation for mathematics.

Other Sciences

Axioms play a key role not only in mathematics, but also in other sciences, notably in theoretical physics. In particular, the monumental work of Isaac Newton is essentially based on Euclid's axioms, augmented by a postulate on the non-relation of spacetime and the physics taking place in it at any moment.

In 1905, Newton's axioms were replaced by those of Albert Einstein's special relativity, and later on by those of general relativity.

Another paper of Albert Einstein and coworkers (see EPR paradox), almost immediately contradicted by Niels Bohr, concerned the interpretation of quantum mechanics. This was in 1935. According to Bohr, this new theory should be probabilistic, whereas according to Einstein it should be deterministic. Notably, the underlying quantum mechanical theory, i.e. the set of "theorems" derived by it, seemed to be identical. Einstein even assumed that it would be sufficient to add to quantum mechanics "hidden variables" to enforce determinism. However, thirty years later, in 1964, John Bell found a theorem, involving complicated optical correlations (see Bell inequalities), which yielded measurably different results using Einstein's axioms compared to using Bohr's axioms. And it took roughly another twenty years until an experiment of Alain Aspect got results in favour of Bohr's axioms, not Einstein's. (Bohr's axioms are simply: The theory should be probabilistic in the sense of the Copenhagen interpretation.)

As a consequence, it is not necessary to explicitly cite Einstein's axioms, the more so since they concern subtle points on the "reality" and "locality" of experiments.

Regardless, the role of axioms in mathematics and in the above-mentioned sciences is different. In mathematics one neither "proves" nor "disproves" an axiom for a set of theorems; the point is simply that in the conceptual realm identified by the axioms, the theorems logically follow. In contrast, in physics a comparison with experiments always makes sense, since a falsified physical theory needs modification.

MATHEMATICAL LOGIC

In the field of mathematical logic, a clear distinction is made between two notions of axioms: *logical* and *non-logical* (somewhat similar to the ancient distinction between "axioms" and "postulates" respectively).

Logical Axioms

These are certain formulas in a formal language that are universally valid, that is, formulas that are satisfied by every assignment of values. Usually one takes as logical axioms *at least* some minimal set of tautologies that is sufficient for proving all tautologies in the language; in the case of predicate logic more logical axioms than that are required, in order to prove logical truths that are not tautologies in the strict sense.

Examples

Propositional Logic

In propositional logic it is common to take as logical axioms all formulae of the following forms, where ϕ, χ, and ψ can be any formulae of the language and where the included primitive connectives are only "$\neg$" for negation of the immediately following proposition and "$\rightarrow$" for implication from antecedent to consequent propositions:

1. $\phi \rightarrow (\psi \rightarrow \phi)$
2. $(\phi \rightarrow (\psi \rightarrow \chi)) \rightarrow ((\phi \rightarrow \psi) \rightarrow (\phi \rightarrow \chi))$
3. $(\neg\phi \rightarrow \neg\psi) \rightarrow (\psi \rightarrow \phi)$.

Each of these patterns is an *axiom schema*, a rule for generating an infinite number of axioms. For example, if AB and C are propositional variables, then $A \rightarrow (B \rightarrow A)$ and $(A \rightarrow \neg B) \rightarrow (C \rightarrow (A \rightarrow \neg B))$ are both instances of axiom schema 1, and hence are axioms. It can be shown that with only these three axiom schemata and *modus ponens*, one can prove all tautologies of the propositional calculus. It can also be shown that no pair of these schemata is sufficient for proving all tautologies with *modus ponens*.

Other axiom schemas involving the same or different sets of primitive connectives can be alternatively constructed.

These axiom schemata are also used in the predicate calculus, but additional logical axioms are needed to include a quantifier in the calculus.

First-Order Logic

Axiom of Equality. Let $\mathfrak{L}$ be a first-order language. For each variable x the formula

x = x

is universally valid.

This means that, for any variable symbol x, the formula x = x, can be regarded as an axiom. Also, in this example, for this not to fall into vagueness and a never-ending series of "primitive notions", either a precise notion of what we mean by x = x (or, for that matter, "to be equal") has to be well established first, or a purely formal and syntactical usage of the symbol = has to be enforced, only regarding it as a string and only a string of symbols, and mathematical logic does indeed do that.

Another, more interesting example axiom scheme, is that which provides us with what is known as Universal Instantiation:

Axiom scheme for Universal Instantiation. Given a formula ϕ in a first-order language $\mathfrak{L}$ a variable x and a term t that is substitutable for x in ϕ the formula

$\forall x\, \phi \to \phi^x_t$

is universally valid.

Where the symbol ϕ^x_t stands for the formula ϕ with the term t substituted for x (See Substitution of variables.) In informal terms, this example allows us to state that, if we know that a certain property P holds for every x and that t stands for a particular object in our structure, then we should be able to claim $P(t)$ Again, we are claiming that the formula $\forall x \phi \to \phi^x_t$ is valid, that is, we must be able to give a "proof" of this fact, or more properly speaking, a metaproof. Actually, these examples are metatheorems *of our theory of mathematical logic since we are dealing with the very concept of proof itself. Aside from this, we can also have* Existential Generalization:

Axiom scheme for Existential Generalization. *Given a formula* ϕ in a first-order language $\mathfrak{L}$ a variable x and a term t that is substitutable for x in ϕ the formula

$\phi^x_t \to \exists x\, \phi$

is universally valid.

Non-Logical Axioms

Non-logical axioms are formulas that play the role of theory-specific assumptions. Reasoning about two different structures, for example the natural numbers and the integers, may involve the same logical axioms; the non-logical axioms aim to capture what is special about a particular structure (or set of structures, such as groups). Thus non-logical axioms, unlike logical axioms, are not *tautologies*. Another name for a non-logical axiom is *postulate*.

Almost every modern mathematical theory starts from a given set of non-logical axioms, and it was thought that in principle every theory could be axiomatized in this way and formalized down to the bare language of logical formulas.

Non-logical axioms are often simply referred to as *axioms* in mathematical discourse. This does not mean that it is claimed that they are true in some absolute sense. For example, in some groups, the group operation is commutative, and this can be asserted with the introduction of an additional axiom, but without this axiom we can do quite well developing (the more general) group theory, and we can even take its negation as an axiom for the study of non-commutative groups.

Thus, an *axiom* is an elementary basis for a formal logic system that together with the rules of inference define a **deductive system**.

Examples

This section gives examples of mathematical theories that are developed entirely from a set of non-logical axioms (axioms, henceforth). A rigorous treatment of any of these topics begins with a specification of these axioms.

Basic theories, such as arithmetic, real analysis and complex analysis are often introduced non-axiomatically, but implicitly or explicitly there is generally an assumption that the axioms being used are the axioms of Zermelo–Fraenkel set theory with choice, abbreviated ZFC, or some very similar system of axiomatic set theory like Von Neumann–Bernays–Gödel set theory, a conservative extension of ZFC. Sometimes slightly stronger theories such as Morse-Kelley set theory or set theory with a strongly inaccessible cardinal allowing the use of a Grothendieck universe are used, but in fact most mathematicians can actually prove all they need in systems weaker than ZFC, such as second-order arithmetic.

The study of topology in mathematics extends all over through point set topology, algebraic topology, differential topology, and all the related paraphernalia, such as homology theory, homotopy theory. The development of *abstract algebra* brought with itself group theory, rings, fields, and Galois theory.

This list could be expanded to include most fields of mathematics, including measure theory, ergodic theory, probability, representation theory, and differential geometry.

Combinatorics is an example of a field of mathematics which does not, in general, follow the axiomatic method.

Arithmetic

The Peano axioms are the most widely used *axiomatization* of first-order arithmetic. They are a set of axioms strong enough to prove many important facts about number theory and they allowed Gödel to establish his famous second incompleteness theorem.

We have a language $\mathfrak{L}_{NT} = \{0, S\}$ where 0 is a constant symbol and S is a unary function and the following axioms:

1. $\forall x. \neg(Sx = 0)$
2. $\forall x. \forall y. (Sx = Sy \to x = y)$
3. $((\phi(0) \land \forall x. (\phi(x) \to \phi(Sx))) \to \forall x. \phi(x)$ for any $\mathfrak{L}_{NT}$ formula ϕ with one free variable.

The standard structure is $\mathfrak{N} = \langle \mathbb{N}, 0, S \rangle$ where $\mathbb{N}$ is the set of natural numbers, S is the successor function and 0 is naturally interpreted as the number 0.

Euclidean Geometry

Probably the oldest, and most famous, list of axioms are the 4 + 1 Euclid's postulates of plane geometry. The axioms are referred to as "4 + 1" because for nearly two millennia the fifth (parallel) postulate ("through a point outside a line there is exactly one parallel") was suspected of being derivable from the first four. Ultimately, the fifth postulate was found to be independent of the first four. Indeed, one can assume that exactly one parallel through a point outside a line exists, or that infinitely many exist. This choice gives us two alternative forms of geometry in which the interior angles of a triangle add up to exactly 180 degrees or less, respectively, and are known as Euclidean and hyperbolic geometries. If one also removes the second postulate ("a line can be extended indefinitely") then elliptic geometry arises, where there is no parallel through a point outside a line, and in which the interior angles of a triangle add up to more than 180 degrees.

Real Analysis

The objectives of study are within the domain of real numbers. The real numbers are uniquely picked out (up to isomorphism) by the properties of a *Dedekind complete ordered field*, meaning that any nonempty set of real numbers with an upper bound has a least upper bound. However, expressing these properties as axioms requires use of second-order logic. The Löwenheim-Skolem theorems tell us that if we restrict ourselves to first-order logic, any axiom system for the reals admits other models, including both models that are smaller than the reals and models that are larger. Some of the latter are studied in non-standard analysis.

Role in Mathematical Logic

Deductive Systems and Completeness

A deductive system consists of a set Λ of logical axioms, a set Σ of non-logical axioms, and a set $\{(\Gamma, \phi)\}$ of *rules of inference*. A desirable property of a deductive system is that it be **complete**. A system is said to be complete if, for all formulas ϕ

$$\text{if } \Sigma \models \phi \text{ then } \Sigma \vdash \phi$$

that is, for any statement that is a *logical consequence* of Σ there actually exists a *deduction* of the statement from Σ This is sometimes expressed as "everything that is true is provable", but it must be understood that "true" here means "made true by the set of axioms", and not, for example, "true in the intended interpretation". Gödel's completeness theorem establishes the completeness of a certain commonly used type of deductive system.

Note that "completeness" has a different meaning here than it does in the context of Gödel's first incompleteness theorem, which states that no *recursive*, *consistent* set of non-logical axioms Σ of the Theory of Arithmetic is *complete*, in the sense that there will always exist an arithmetic statement ϕ, such that neither ϕ nor $\neg\phi$ can be proved from the given set of axioms.

There is thus, on the one hand, the notion of *completeness of a deductive system* and on the other hand that of *completeness of a set of non-logical axioms*. The completeness theorem and the incompleteness theorem, despite their names, do not contradict one another.

REFERENCES

1. Cf. axiom, n., etymology. Oxford English Dictionary, accessed 2012-04-28.
2. Oxford American College Dictionary: "n. a statement or proposition that is regarded as being established, accepted, or self-evidently true. ORIGIN: late 15th cent.: ultimately from Greek axiōma 'what is thought fitting,' from axios 'worthy.' http://www.highbeam.com/doc/1O997-axiom.html (subscription required)
3. "A proposition that commends itself to general acceptance; a well-established or universally conceded principle; a maxim, rule, law" axiom, n., definition 1a. Oxford English Dictionary Online, accessed 2012-04-28. Cf. Aristotle, Posterior Analytics I.2.72a18-b4.
4. "A proposition (whether true or false)" axiom, n., definition 2. Oxford English Dictionary Online, accessed 2012-04-28.
5. Maddy, Penelope (Jun 1988). "Believing the Axioms, I". Journal of Symbolic Logic. 53 (2): 481–511. doi:10.2307/2274520. for a realist view.
6. Wolff, P. Breakthroughs in Mathematics, 1963, New York: New American Library, pp 47–48
7. Heath, T. 1956. The Thirteen Books of Euclid's Elements. New York: Dover. p 200
8. Aristotle, Metaphysics Bk IV, Chapter 3, 1005b "Physics also is a kind of Wisdom, but it is not the first kind. – And the attempts of some of those who discuss the terms on which truth should be accepted, are due to want of training in logic; for they should know these things already when they come to a special

study, and not be inquiring into them while they are listening to lectures on it." W.D. Ross translation, in The Basic Works of Aristotle, ed. Richard McKeon, (Random House, New York, 1941)|date=June 2011

9. Mendelson, "6. Other Axiomatizations" of Ch. 1
10. Mendelson, "3. First-Order Theories" of Ch. 2
11. Mendelson, "3. First-Order Theories: Proper Axioms" of Ch. 2
12. Mendelson, "5. The Fixed Point Theorem. Gödel's Incompleteness Theorem" of Ch. 2

CHAPTER 5

Conjecture

Chapter Contents

INTRODUCTION

In mathematics, a conjecture is a conclusion or proposition based on incomplete information, for which no proof has been found. Conjectures such as the Riemann hypothesis (still a conjecture) or Fermat's Last Theorem (which was a conjecture until proven in 1995) have shaped much of mathematical history as new areas of mathematics are developed in order to prove them.

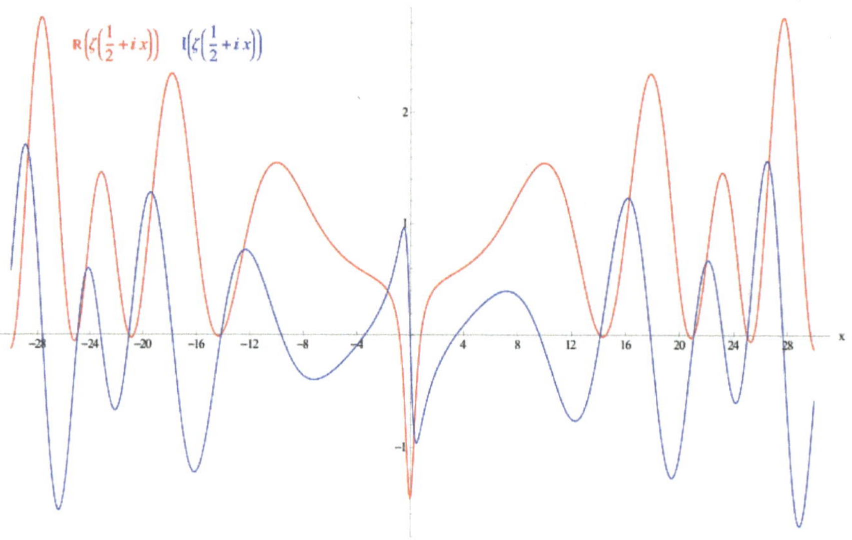

The real part (red) and imaginary part (blue) of the Riemann zeta function along the critical line Re(s) = 1/2. The first non-trivial zeros can be seen at Im(s) = ±14.135, ±21.022 and ±25.011. The Riemann hypothesis, a famous conjecture, says that all non-trivial zeros of the zeta function lie along the critical line.

FERMAT'S LAST THEOREM

In number theory, Fermat's Last Theorem (sometimes called Fermat's conjecture, especially in older texts) states that no three positive integers a, b, and c satisfy the equation $a^n + b^n = c^n$ for any integer value of n greater than 2. The cases n = 1 and n = 2 have been known to have infinitely many solutions since antiquity.

This theorem was first conjectured by Pierre de Fermat in 1637 in the margin of a copy of Arithmetica where he claimed he had a proof that was too large to fit in the margin. The first successful proof was released in 1994 by Andrew Wiles, and formally published in 1995, after 358 years of effort by mathematicians. The unsolved problem stimulated the development of algebraic number theory in the 19th century and the proof of the modularity theorem in the 20th century. It is among the most notable theorems in the history of mathematics and prior to its proof, it was in the Guinness Book of World Records as the "most difficult mathematical problem", one of the reasons being that it has the largest number of unsuccessful proofs.

Overview

The Pythagorean equation, $x^2 + y^2 = z^2$, has an infinite number of positive integer solutions for x, y, and z; these solutions are known as Pythagorean triples. Around 1637, Fermat wrote in the margin of a book that the more general equation $a^n + b^n = c^n$ had no solutions in positive integers, if n is an integer greater than 2. Although he claimed to have a general proof of his conjecture, Fermat left no details of his proof, and no proof by him has ever been found. His claim was discovered some 30 years later, after his death. This claim, which came to be known as Fermat's Last Theorem, stood unsolved in mathematics for the following three and a half centuries.

The claim eventually became one of the most notable unsolved problems of mathematics. Attempts to prove it prompted substantial development in number theory, and over time Fermat's Last Theorem gained prominence as an unsolved problem in mathematics.

Subsequent Developments and Solution

With the special case n = 4 proved, it suffices to prove the theorem for exponents n that are prime numbers (this reduction is considered trivial to prove[1]). Over the next two centuries (1637–1839), the conjecture was proved for only the primes 3, 5, and 7, although Sophie Germain innovated and proved an approach that was relevant to an entire class of primes. In the mid-19th century, Ernst Kummer extended this and proved the theorem for all regular primes, leaving irregular primes to be analyzed individually. Building on Kummer's work and using sophisticated computer studies, other mathematicians were able to extend the proof to cover all prime exponents up to four million, but a proof for all exponents was inaccessible (meaning that mathematicians generally considered a proof impossible, exceedingly difficult, or unachievable with current knowledge).

Around 1955, Japanese mathematicians Goro Shimura and Yutaka Taniyama suspected a link might exist between elliptic curves and modular forms, two completely different areas of mathematics. Known at the time as the Taniyama–Shimura-Weil conjecture, and (eventually) as the modularity theorem, it stood on its own, with no apparent connection to Fermat's Last Theorem. It was widely seen as significant and important in its own right, but was (like Fermat's theorem) widely considered completely inaccessible to proof.

In 1984, Gerhard Frey noticed an apparent link between the modularity theorem and Fermat's Last Theorem. This potential link was confirmed two years later by Ken Ribet, who gave a conditional proof of Fermat's Last Theorem that depended on the modularity theorem (see: Ribet's Theorem and Frey curve). On hearing this, English mathematician Andrew Wiles, who had a childhood fascination with Fermat's Last Theorem, decided to try to prove the modularity theorem as a way to prove Fermat's Last Theorem. In 1993, after six years working secretly on the problem, Wiles succeeded in proving enough of the modularity theorem to prove Fermat's Last Theorem for odd prime exponents. Wiles's paper was massive in size and scope. A flaw was discovered in one part of his original paper during peer review and required a further year and collaboration with a past student, Richard Taylor, to resolve. As a result, the final proof in 1995 was accompanied by a second, smaller, joint paper to that effect. Wiles's achievement was reported widely in the popular press, and was popularized in books and television programs. The remaining parts of the modularity theorem were subsequently proved by other mathematicians, building on Wiles's work, between 1996 and 2001.. For his proof, Wiles was honoured and received numerous awards, including the 2016 Abel Prize.

Equivalent Statements of the Theorem

There are several simple alternative ways to state Fermat's Last Theorem that are equivalent to the one given above. In order to state them, let **N** be the set of natural numbers 1,2,3,..., let **Z** be the set of integers 0, ±1, ±2,..., and let **Q** be the set of rational numbers a/b where a and bare in **Z** with $b \neq 0$.

In what follows we will call a solution to $x^n + y^n = z^n$ where one or more of x, y, or z is zero a trivial solution. A solution where all three are non-zero will be called a non-trivial solution. For comparison's sake we start with the original formulation.

Original statement. With n, x, y, z ∈ **N** and $n > 2$ the equation $x^n + y^n = z^n$ has no solutions.

Most popular domain treatments of the subject state it this way. In contrast, almost all math textbooks state it over **Z**:

Equivalent statement 1: $x^n + y^n = z^n$, where $n \geq 3$, has no non-trivial solutions x, y, z ∈ **Z**.

The equivalence is clear if n is even. If n is odd and all three of x, y, z are negative then we can replace x, y, z with −x, −y, −z to obtain a solution in **N**. If two of them are negative, it must be x and z or y and z. If x, z are negative and y is positive, then we can rearrange to get $(-z)^n + y^n = (-x)^n$resulting in a solution in **N**; the other case is dealt with analogously. Now if just one is negative, it must be x or y. If x is negative, and y and z are positive, then it can be rearranged to get $(-x)^n + z^n = y^n$ again resulting in a solution in **N**; if y is negative, the result follows symmetrically. Thus in all cases a nontrivial solution in **Z** results in a solution in **N**.

Equivalent statement 2: $x^n + y^n = z^n$, where $n \geq 3$, has no non-trivial solutions x, y, z ∈ **Q**.

This is because the exponent of x, y and z are equal (to n), so if there is a solution in **Q** then it can be multiplied through by an appropriate common denominator to get a solution in **Z**, and hence in **N**.

Equivalent statement 3: $x^n + y^n = 1$, where $n \geq 3$, has no non-trivial solutions x, y ∈ **Q**.

A non-trivial solution a, b, c ∈ **Z** to $x^n + y^n = z^n$ yields the non-trivial solution a/c, b/c ∈ **Q** for $v^n + w^n = 1$. Conversely, a solution a/b, c/d ∈ **Q** to $v^n + w^n = 1$yields the non-trivial solution ad, cb, bd for $x^n + y^n = z^n$.

This last formulation is particularly fruitful, because it reduces the problem from a problem about surfaces in three dimensions to a problem about curves in two dimensions. Furthermore, it allows working over the field **Q**, rather than over the ring **Z**; fields exhibit more structure than rings, which allows for deeper analysis of their elements.

Connection to elliptic curves: If a, b, c is a non-trivial solution to $x^p + y^p = z^p$, p odd prime, then $y^2 = x(x - a^p)(x + b^p)$ (Frey curve) is an elliptic curve.

Examining this elliptic curve with Ribet's theorem shows that it cannot have a modular form. The proof by Andrew Wiles shows that $y^2 = x(x - a^n)(x + b^n)$ always has a modular form. This implies that a non-trivial solution to $x^p + y^p = z^p$, p odd prime, would create a contradiction. This shows that no non-trivial solutions exist.

MATHEMATICAL HISTORY

Pythagoras and Diophantus

Pythagorean Triples

A Pythagorean triple – named for the ancient Greek Pythagoras – is a set of three integers (a, b, c) that satisfy a special case of Fermat's equation (n = 2)

$$a^2 + b^2 = c^2.$$

Examples of Pythagorean triples include (3, 4, 5) and (5, 12, 13). There are infinitely many such triples, and methods for generating such triples have been studied in many cultures, beginning with the Babylonians and later ancient Greek, Chinese, and Indian mathematicians. The traditional interest in Pythagorean triples connects with the Pythagorean theorem; in its converse form, it states that a triangle with sides of lengths a, b, and c has a right angle between the a and b legs when the numbers are a Pythagorean triple. Fermat's Last Theorem is an extension of this problem to higher powers, stating that no solution exists when the exponent 2 is replaced by any larger integer.

Diophantine Equations

Fermat's equation, $x^n + y^n = z^n$ with positive integer solutions, is an example of a Diophantine equation, named for the 3rd-century Alexandrian mathematician, Diophantus, who studied them and developed methods for the solution of some kinds of Diophantine equations. A typical Diophantine problem is to find two integers x and y such that their sum, and the sum of their squares, equal two given numbers A and B, respectively:

$$A = x + y$$
$$B = x^2 + y^2.$$

Diophantus's major work is the Arithmetica, of which only a portion has survived. Fermat's conjecture of his Last Theorem was inspired while reading a new edition of the Arithmetica, that was translated into Latin and published in 1621 by Claude Bachet.

Diophantine equations have been studied for thousands of years. For example, the solutions to the quadratic Diophantine equation $x^2 + y^2 = z^2$are given by the Pythagorean triples, originally solved by the Babylonians (c. 1800 BC). Solutions to linear Diophantine equations, such as $26x + 65y = 13$, may be found using the Euclidean algorithm (c. 5th century BC). Many Diophantine equations have a form similar to the equation of Fermat's Last Theorem from the point of view of algebra, in that they have no cross terms mixing two letters, without sharing its particular properties. For example, it is known that there are infinitely many positive integers x, y, and z such that $x^n + y^n = z^m$ where n and m are relatively prime natural numbers.

Fermat's Conjecture

Problem II.8 of the Arithmetica asks how a given square number is split into two other squares; in other words, for a given rational number k, find rational numbers u and v such that $k^2 = u^2 + v^2$. Diophantus shows how to solve this sum-of-squares problem for $k = 4$ (the solutions being $u = 16/5$ and $v = 12/5$).

Around 1637, Fermat wrote his Last Theorem in the margin of his copy of the Arithmetica next to Diophantus' sum-of-squares problem:

Cubum autem in duos cubos, aut quadratoquadratum in duos quadratoquadratos & generaliter nullam in infinitum ultra quadratum potestatem in duos eiusdem nominis fas est dividere cuius rei demonstrationem mirabilem sane detexi. Hanc marginis exiguitas non caperet.	It is impossible to separate a cube into two cubes, or a fourth power into two fourth powers, or in general, any power higher than the second, into two like powers. I have discovered a truly marvelous proof of this, which this margin is too narrow to contain.

After Fermat's death in 1665, his son Clément-Samuel Fermat produced a new edition of the book (1670) augmented with his father's comments. The margin note became known as Fermat's Last Theorem,as it was the last of Fermat's asserted theorems to remain unproved.

It is not known whether Fermat had actually found a valid proof for all exponents n, but it appears unlikely. Only one related proof by him has survived, namely for the case $n = 4$, as described in the section Proofs for specific exponents. While Fermat posed the cases of $n = 4$ and of $n = 3$ as challenges to his mathematical correspondents, such as Marin Mersenne, Blaise Pascal, and John Wallis, he never posed the general case. Moreover, in the last thirty years of his life, Fermat never again wrote of his "truly marvelous proof" of the general case, and never published it. Van der Poorten suggests that while the absence of a proof is insignificant, the lack of challenges means Fermat realised he did not have a proof; he quotes Weil as saying Fermat must have briefly deluded himself with an irretrievable idea.

The techniques Fermat might have used in such a marvelous proof are unknown.

Taylor and Wiles's proof relies on 20th-century techniques. Fermat's proof would have had to be elementary by comparison, given the mathematical knowledge of his time.

While Harvey Friedman's grand conjecture implies that any provable theorem (including Fermat's last theorem) can be proved using only 'elementary function arithmetic', such a proof need be 'elementary' only in a technical sense and could involve millions of steps, and thus be far too long to have been Fermat's proof.

Proofs for Specific Exponents

Only one relevant proof by Fermat has survived, in which he uses the technique of infinite descent to show that the area of a right triangle with integer sides can never equal the square of an integer. His proof is equivalent to demonstrating that the equation

$$x^4 - y^4 = z^2$$

has no primitive solutions in integers (no pairwise coprime solutions). In turn, this proves Fermat's Last Theorem for the case n = 4, since the equation $a^4 + b^4 = c^4$ can be written as $c^4 - b^4 = (a^2)^2$.

Alternative proofs of the case n = 4 were developed later by Frénicle de Bessy (1676), Leonhard Euler (1738), Kausler (1802), Peter Barlow (1811), Adrien-Marie Legendre (1830), Schopis (1825), Terquem (1846), Joseph Bertrand (1851), Victor Lebesgue (1853, 1859, 1862),Theophile Pepin (1883), Tafelmacher (1893), David Hilbert (1897), Bendz (1901), Gambioli (1901), Leopold Kronecker (1901), Bang (1905), Sommer (1907), Bottari (1908), Karel Rychlík (1910), Nutzhorn (1912), Robert Carmichael (1913), Hancock (1931), and Vrănceanu (1966).

For another proof for n=4 by infinite descent, see Infinite descent: Non-solvability of $r^2 + s^4 = t^4$. For various proofs for n=4 by infinite descent, see Grant and Perella (1999), Barbara (2007), and Dolan (2011).

After Fermat proved the special case n = 4, the general proof for all n required only that the theorem be established for all odd prime exponents. In other words, it was necessary to prove only that the equation $a^n + b^n = c^n$ has no integer solutions (a, b, c) when n is an odd prime number. This follows because a solution (a, b, c) for a given n is equivalent to a solution for all the factors of n. For illustration, let n be factored into d and e, n = de. The general equation

$$a^n + b^n = c^n$$

implies that (a^d, b^d, c^d) is a solution for the exponent e

$$(a^d)^e + (b^d)^e = (c^d)^e.$$

Thus, to prove that Fermat's equation has no solutions for n > 2, it would suffice to prove that it has no solutions for at least one prime factor of every n. Each integer n > 2 is divisible by 4 or an odd prime number (or both). Therefore, Fermat's Last Theorem could be proved for all n if it could be proved for n = 4 and for all odd primes p.

In the two centuries following its conjecture (1637–1839), Fermat's Last Theorem was proved for three odd prime exponents p = 3, 5 and 7. The case p = 3 was first stated by Abu-Mahmud Khojandi (10th century), but his attempted proof of the theorem was incorrect. In 1770, Leonhard Euler gave a proof of p = 3, but his proof by infinite descent contained a major gap. However, since Euler himself had proved the lemma necessary to complete the proof in other work, he is generally credited with the first proof. Independent proofs were published by Kausler (1802), Legendre (1823, 1830), Calzolari (1855), Gabriel Lamé (1865), Peter Guthrie Tait (1872), Günther (1878), Gambioli (1901), Krey (1909), Rychlík (1910), Stockhaus (1910), Carmichael (1915), Johannes van der Corput (1915), Axel Thue (1917), and Duarte (1944). The case p = 5 was proved independently by Legendre and Peter Gustav Lejeune Dirichlet around 1825.Alternative proofs were developed by Carl Friedrich Gauss (1875, posthumous), Lebesgue (1843), Lamé (1847), Gambioli (1901), Werebrusow (1905), Rychlík (1910), van der Corput (1915), and Guy Terjanian (1987). The case p = 7 was proved by Lamé in 1839. His rather complicated proof was simplified in 1840 by Lebesgue, and still simpler proofs were published by Angelo Genocchi in 1864, 1874 and 1876. Alternative proofs were developed by Théophile Pépin (1876) and Edmond Maillet (1897).

Fermat's Last Theorem was also proved for the exponents n = 6, 10, and 14. Proofs for n = 6 were published by Kausler, Thue,Tafelmacher, Lind, Kapferer, Swift, and Breusch. Similarly, Dirichlet and Terjanian each proved the case n = 14, while Kapferer and Breusch each proved the case n = 10. Strictly speaking, these proofs are unnecessary, since these cases follow from the proofs for n = 3, 5, and 7, respectively. Nevertheless, the reasoning of these even-exponent proofs differs from their odd-exponent counterparts. Dirichlet's proof for n = 14 was published in 1832, before Lamé's 1839 proof for n = 7.

All proofs for specific exponents used Fermat's technique of infinite descent, either in its original form, or in the form of descent on elliptic curves or abelian varieties. The details and auxiliary arguments, however, were often ad hoc and tied to the individual exponent under consideration. Since they became ever more complicated as p increased, it seemed unlikely that the general case of Fermat's Last Theorem could be proved by building upon the proofs for individual exponents. Although some general results on Fermat's Last Theorem were published in the early 19th century by Niels Henrik Abel and Peter Barlow, the first significant work on the general theorem was done by Sophie Germain.

Sophie Germain

In the early 19th century, Sophie Germain developed several novel approaches to prove Fermat's Last Theorem for all exponents. First, she defined a set of auxiliary primes θ constructed from the prime exponent p by the equation $\theta = 2hp + 1$, where h is any integer not divisible by three. She showed that, if no integers raised to the pth power were adjacent modulo θ (the non-consecutivity condition), then θ must divide the product xyz. Her goal was to use mathematical induction to prove that, for any given p, infinitely many auxiliary primes θ satisfied the non-consecutivity condition and thus divided xyz; since the product xyz can have at most a finite number of prime factors, such a proof would have established Fermat's Last Theorem. Although she developed many techniques for establishing the non-consecutivity condition, she did not succeed in her strategic goal. She also worked to set lower limits on the size of solutions to Fermat's equation for a given exponent p, a modified version of which was published by Adrien-Marie Legendre. As a byproduct of this latter work, she proved Sophie Germain's theorem, which verified the first case of Fermat's Last Theorem (namely, the case in which p does not divide xyz) for every odd prime exponent less than 100. Germain tried unsuccessfully to prove the first case of Fermat's Last Theorem for all even exponents, specifically for $n = 2p$, which was proved by Guy Terjanian in 1977. In 1985, Leonard Adleman, Roger Heath-Brown and Étienne Fouvry proved that the first case of Fermat's Last Theorem holds for infinitely many odd primes p.

Ernst Kummer and the Theory of Ideals

In 1847, Gabriel Lamé outlined a proof of Fermat's Last Theorem based on factoring the equation $x^p + y^p = z^p$ in complex numbers, specifically the cyclotomic field based on the roots of the number 1. His proof failed, however, because it assumed incorrectly that such complex numbers can be factored uniquely into primes, similar to integers. This gap was pointed out immediately by Joseph Liouville, who later read a paper that demonstrated this failure of unique factorisation, written by Ernst Kummer.

Kummer set himself the task of determining whether the cyclotomic field could be generalized to include new prime numbers such that unique factorisation was restored. He succeeded in that task by developing the ideal numbers. Using the general approach outlined by Lamé, Kummer proved both cases of Fermat's Last Theorem for all regular prime numbers. However, he could not prove the theorem for the exceptional primes (irregular primes) that conjecturally occur approximately 39% of the time; the only irregular primes below 100 are 37, 59 and 67.

Mordell Conjecture

In the 1920s, Louis Mordell posed a conjecture that implied that Fermat's equation has at most a finite number of nontrivial primitive integer solutions, if the exponent n is greater than two. This conjecture was proved in 1983 by Gerd Faltings, and is now known as Faltings' theorem.

Computational Studies

In the latter half of the 20th century, computational methods were used to extend Kummer's approach to the irregular primes. In 1954, Harry Vandiver used a SWAC computer to prove Fermat's Last Theorem for all primes up to 2521. By 1978, Samuel Wagstaff had extended this to all primes less than 125,000. By 1993, Fermat's Last Theorem had been proved for all primes less than four million.

However despite these efforts and their results, no proof existed of Fermat's Last Theorem. Proofs of individual exponents by their nature could never prove the general case: even if all exponents were verified up to an extremely large number X, a higher exponent beyond X might still exist for which the claim was not true. (This had been the case with some other past conjectures, and it could not be ruled out in this conjecture.)

Connection with Elliptic Curves

The strategy that ultimately led to a successful proof of Fermat's Last Theorem arose from the "astounding"[:211] Taniyama–Shimura-Weil conjecture, proposed around 1955—which many mathematicians believed would be near to impossible to prove,[:223] and was linked in the 1980s by Gerhard Frey, Jean-Pierre Serre and Ken Ribet to Fermat's equation. By accomplishing a partial proof of this conjecture in 1994, Andrew Wiles ultimately succeeded in proving Fermat's Last Theorem, as well as leading the way to a full proof by others of what is now the modularity theorem.

Taniyama–Shimura–Weil Conjecture

Around 1955, Japanese mathematicians Goro Shimura and Yutaka Taniyama observed a possible link between two apparently completely distinct branches of mathematics, elliptic curves and modular forms. The resulting modularity theorem (at the time known as the Taniyama–Shimura conjecture) states that every elliptic curve is modular, meaning that it can be associated with a unique modular form.

It was initially dismissed as unlikely or highly speculative, and was taken more seriously when number theorist André Weil found evidence supporting it, but no proof; as a result the conjecture was often known as the Taniyama–Shimura-Weil conjecture. It became a part of the Langlands programme, a list of important conjectures needing proof or disproof.

Even after gaining serious attention, the conjecture was seen by contemporary mathematicians as extraordinarily difficult or perhaps inaccessible to proof. For example, Wiles's ex-supervisor John Coates states that it seemed "impossible to actually prove", and Ken Ribet considered himself "one of the vast majority of people who believed it was completely inaccessible, adding that Andrew Wiles was probably one of the few people on earth who had the audacity to dream that you can actually go and prove it.

Ribet's Theorem for Frey Curves

Main articles: Frey curve and Ribet's theorem

In 1984, Gerhard Frey noted a link between Fermat's equation and the modularity theorem, then still a conjecture. If Fermat's equation had any solution (a, b, c) for exponent $p > 2$, then it could be shown that the elliptic curve (now known as a Frey-Hellegouarch)

$y^2 = x (x - a^p)(x + b^p)$

would have such unusual properties that it was unlikely to be modular. This would conflict with the modularity theorem, which asserted that all elliptic curves are modular. As such, Frey observed that a proof of the Taniyama–Shimura-Weil conjecture would simultaneously prove Fermat's Last Theorem and equally, a disproof or refutation of Fermat's Last Theorem would disprove the conjecture.

Following this strategy, a proof of Fermat's Last Theorem required two steps. First, it was necessary to prove the modularity theorem – or at least to prove it for the subclass of cases (known as semistable elliptic curves) that included Frey's equation – and this was widely believed inaccessible to proof by contemporary mathematicians. Second, it was necessary to show that Frey's intuition was correct: that if an elliptic curve were constructed in this way, using a set of numbers that were a solution of Fermat's equation, the resulting elliptic curve could not be modular. Frey did not quite succeed in proving this rigorously; the missing piece (the so-called "epsilon conjecture", now known as Ribet's theorem) was identified by Jean-Pierre Serre and proved in 1986 by Ken Ribet.

- The modularity theorem – if proved – would mean all elliptic curves (or at least all semistable elliptic curves) are of necessity modular.
- Ribet's theorem – proved in 1986 – showed that, if a solution to Fermat's equation existed, it could be used to create a semistable elliptic curve that was not modular;
- The contradiction would imply (if the modularity theorem were correct) that no solutions can exist to Fermat's equation – therefore proving Fermat's Last Theorem.

Wiles's General Proof

British mathematician Andrew Wiles.

Ribet's proof of the epsilon conjecture in 1986 accomplished the first of the two goals proposed by Frey. Upon hearing of Ribet's success, Andrew Wiles, an English mathematician with a childhood fascination with Fermat's Last Theorem, and a prior study area of elliptical equations, decided to commit himself to accomplishing the second half: proving a special case of the modularity theorem (then known as the Taniyama–Shimura conjecture) for semistable elliptic curves.

Wiles worked on that task for six years in near-total secrecy, covering up his efforts by releasing prior work in small segments as separate papers and confiding only in his wife. His initial study suggested proof by induction, and he based his initial work and first significant breakthrough on Galois theory before switching to an attempt to extend horizontal Iwasawa theory for the inductive argument around 1990–91 when it seemed that there was no existing approach adequate to the problem. However, by the summer of 1991, Iwasawa theory also seemed to not be reaching the central issues in the problem. In response, he approached colleagues to seek out any hints of cutting edge research and new techniques, and discovered an Euler system recently developed by Victor Kolyvagin and Matthias Flach that seemed "tailor made" for the inductive part of his proof. Wiles studied and extended this approach, which worked. Since his work relied extensively on this approach, which was new to mathematics and to Wiles, in January 1993 he asked his Princeton colleague, Nick Katz, to check his reasoning for subtle errors. Their conclusion at the time was that the techniques Wiles used seemed to work correctly.

By mid-May 1993, Wiles felt able to tell his wife he thought he had solved the proof of Fermat's Last Theorem, and by June he felt sufficiently confident to present his results in three lectures delivered on 21–23 June 1993 at the Isaac Newton Institute for Mathematical Sciences. Specifically, Wiles presented his proof of the Taniyama–Shimura conjecture for semistable elliptic curves; together with Ribet's proof of the epsilon conjecture, this implied Fermat's Last Theorem. However, it became apparent during peer review that a critical point in the proof was incorrect. It contained an error in a bound on the order of a particular group. The error was caught by several mathematicians refereeing Wiles's manuscript including Katz (in his role as reviewer), who alerted Wiles on 23 August 1993.

The error would not have rendered his work worthless – each part of Wiles's work was highly significant and innovative by itself, as were the many developments and techniques he had created in the course of his work, and only one part was affected. However without this part proved, there was no actual proof of Fermat's Last Theorem. Wiles spent almost a year trying to repair his proof, initially by himself and then in collaboration with Richard Taylor, without success.

On 19 September 1994, on the verge of giving up, Wiles had a flash of insight that the proof could be saved by returning to his original Horizontal Iwasawa theory approach, which he had abandoned in favour of the Kolyvagin–Flach approach, this time strengthening it with expertise gained in Kolyvagin–Flach's approach. On 24 October 1994, Wiles submitted two manuscripts, Modular elliptic curves and Fermat's Last Theoremand Ring theoretic properties of certain Hecke algebras, the second of which was co-authored with Taylor and proved that certain conditions were met that were needed to justify the corrected step in the main paper. The two papers were vetted and published as the entirety of the May 1995 issue of the Annals of Mathematics. These papers established the modularity theorem for semistable elliptic curves, the last step in proving Fermat's Last Theorem, 358 years after it was conjectured.

Subsequent Developments

The full Taniyama–Shimura–Weil conjecture was finally proved by Diamond (1996), Conrad, Diamond & Taylor (1999), and Breuil et al. (2001) who, building on Wiles's work, incrementally chipped away at the remaining cases until the full result was proved. The now fully proved conjecture became known as the modularity theorem.

Several other theorems in number theory similar to Fermat's Last Theorem also follow from the same reasoning, using the modularity theorem. For example: no cube can be written as a sum of two coprime n-th powers, $n \geq 3$. (The case $n = 3$ was already known by Euler.)

Exponents Other than Positive Integers

Reciprocal Integers (Inverse Fermat Equation)

The equation $a^{1/m} + b^{1/m} = c^{1/m}$ can be considered the "inverse" Fermat equation. All solutions of this equation were computed by Lenstra in 1992. In the case in which the mth roots are required to be real and positive, all solutions are given by

$$a = rs^m$$
$$b = rt^m$$
$$c = r(s + t)^m$$

for positive integers r, s, t with s and t coprime.

Rational Exponents

For the Diophantine equation $a^{n/m} + b^{n/m} = c^{n/m}$ with n not equal to 1, Bennett, Glass, and Székely proved in 2004 for $n > 2$, that if n and m are coprime, then there are integer solutions if and only if 6 divides m, and $a^{1/m}$, $b^{1/m}$, and $c^{1/m}$ are different complex 6th roots of the same real number.

Negative Exponents

$n = -1$

All primitive (pairwise coprime) integer solutions to the optic equation $a^{-1} + b^{-1} = c^{-1}$ can be written as

$$a = mn + m^2,$$
$$b = mn + n^2,$$
$$c = mn$$

for positive, coprime integers m, n.

n = –2

The case n = –2 also has an infinitude of solutions, and these have a geometric interpretation in terms of right triangles with integer sides and an integer altitude to the hypotenuse. All primitive solutions to $a^{-2} + b^{-2} = d^{-2}$ are given by

$$a = (v^2 - u^2)(v^2 + u^2),$$
$$b = 2uv(v^2 + u^2),$$
$$d = 2uv(v^2 - u^2),$$

for coprime integers u, v with v > u. The geometric interpretation is that a and b are the integer legs of a right triangle and d is the integer altitude to the hypotenuse. Then the hypotenuse itself is the integer

$$c = (v^2 + u^2)^2,$$

so (a, b, c) is a Pythagorean triple.

Integer *n* < –2

There are no solutions in integers for $a^n + b^n = c^n$ for integers n < –2. If there were, the equation could be multiplied through by $a^{|n|}b^{|n|}c^{|n|}$ to obtain $(bc)^{|n|} + (ac)^{|n|} = (ab)^{|n|}$ which is impossible by Fermat's Last Theorem.

Base Values other than Positive Integers

Fermat's last theorem can easily be extended to positive rationals:

$$\left(\frac{a}{x}\right)^n + \left(\frac{b}{y}\right)^n = \left(\frac{c}{z}\right)^n$$

can have no solutions with n > 2, because any solution could be rearranged as:

$$(ayz)^n + (bxz)^n = (cxy)^n,$$

to which Fermat's Last Theorem applies.

Monetary Prizes

In 1816, and again in 1850, the French Academy of Sciences offered a prize for a general proof of Fermat's Last Theorem. In 1857, the Academy awarded 3000 francs and a gold medal to Kummer for his research on ideal numbers, although he had not submitted an entry for the prize. Another prize was offered in 1883 by the Academy of Brussels.

In 1908, the German industrialist and amateur mathematician Paul Wolfskehl bequeathed 100,000 gold marks—a large sum at the time—to the Göttingen Academy of Sciences to offer as a prize for a complete proof of Fermat's Last Theorem. On 27 June 1908, the

Academy published nine rules for awarding the prize. Among other things, these rules required that the proof be published in a peer-reviewed journal; the prize would not be awarded until two years after the publication; and that no prize would be given after 13 September 2007, roughly a century after the competition was begun. Wiles collected the Wolfskehl prize money, then worth \$50,000, on 27 June 1997. In March 2016, Wiles was awarded the Norwegian government's Abel prize worth €600,000 for "for his stunning proof of Fermat's Last Theorem by way of the modularity conjecture for semistable elliptic curves, opening a new era in number theory."

Prior to Wiles's proof, thousands of incorrect proofs were submitted to the Wolfskehl committee, amounting to roughly 10 feet (3 meters) of correspondence. In the first year alone (1907–1908), 621 attempted proofs were submitted, although by the 1970s, the rate of submission had decreased to roughly 3–4 attempted proofs per month. According to F. Schlichting, a Wolfskehl reviewer, most of the proofs were based on elementary methods taught in schools, and often submitted by "people with a technical education but a failed career". In the words of mathematical historian Howard Eves, "Fermat's Last Theorem has the peculiar distinction of being the mathematical problem for which the greatest number of incorrect proofs have been published."

WEIL CONJECTURES

In mathematics, the **Weil conjectures** were some highly influential proposals by André Weil (1949) on the generating functions (known as local zeta-functions) derived from counting the number of points on algebraic varieties over finite fields.

A variety V over a finite field with q elements has a finite number of rational points, as well as points over every finite field with q^k elements containing that field. The generating function has coefficients derived from the numbers N_k of points over the (essentially unique) field with q^kelements.

Weil conjectured that such zeta-functions should be rational functions, should satisfy a form of functional equation, and should have their zeroes in restricted places. The last two parts were quite consciously modeled on the Riemann zeta function and Riemann hypothesis. The rationality was proved by Dwork (1960), the functional equation by Grothendieck (1965), and the analogue of the Riemann hypothesis was proved by Deligne (1974).

Background and History

The earliest antecedent of the Weil conjectures is by Carl Friedrich Gauss and appears in section VII of his Disquisitiones Arithmeticae (Mazur 1974), concerned with roots of unity and Gaussian periods. In article 358, he moves on from the periods that build up towers of quadratic extensions, for the construction of regular polygons; and assumes that p is a prime number such that p − 1 is divisible by 3. Then there is a cyclic cubic field inside the cyclotomic field of pth roots of unity, and a normal integral basis of periods for the integers of this field (an instance of the Hilbert–Speiser theorem). Gauss constructs the order-3 periods, corresponding to the cyclic group $(\mathbf{Z}/p\mathbf{Z})^\times$ of non-zero residues modulo p under multiplication and its unique subgroup of index three. Gauss lets

$\mathfrak{R}$, $\mathfrak{R}'$, and $\mathfrak{R}''$ be its cosets. Taking the periods (sums of roots of unity) corresponding to these cosets applied to exp(2πi/p), he notes that these periods have a multiplication table that is accessible to calculation. Products are linear combinations of the periods, and he determines the coefficients. He sets, for example, $(\mathfrak{R}\mathfrak{R})$ equal to the number of elements of **Z**/p**Z** which are in $\mathfrak{R}$ and which, after being increased by one, are also in $\mathfrak{R}$ He proves that this number and related ones are the coefficients of the products of the periods. To see the relation of these sets to the Weil conjectures, notice that if α and $\alpha + 1$ are both in $\mathfrak{R}$ then there exist x and y in **Z**/p**Z** such that $x^3 = \alpha$ and $y^3 = \alpha + 1$; consequently, $x^3 + 1 = y^3$. Therefore $(\mathfrak{R}\mathfrak{R})$ is the number of solutions to $x^3 + 1 = y^3$ in the finite field **Z**/p**Z**. The other coefficients have similar interpretations. Gauss's determination of the coefficients of the products of the periods therefore counts the number of points on these elliptic curves, and as a byproduct he proves the analog of the Riemann hypothesis.

The Weil conjectures in the special case of algebraic curves were conjectured by Artin (1924). The case of curves over finite fields was proved by Weil, finishing the project started by Hasse's theorem on elliptic curves over finite fields. Their interest was obvious enough from within number theory: they implied upper bounds for exponential sums, a basic concern in analytic number theory (Moreno 2001).

What was really eye-catching, from the point of view of other mathematical areas, was the proposed connection with algebraic topology. Given that finite fields are discrete in nature, and topology speaks only about the continuous, the detailed formulation of Weil (based on working out some examples) was striking and novel. It suggested that geometry over finite fields should fit into well-known patterns relating to Betti numbers, the Lefschetz fixed-point theorem and so on.

The analogy with topology suggested that a new homological theory be set up applying within algebraic geometry. This took two decades (it was a central aim of the work and school of Alexander Grothendieck) building up on initial suggestions from Serre. The rationality part of the conjectures was proved first by Bernard Dwork (1960), using p-adic methods. Grothendieck (1965) and his collaborators established the rationality conjecture, the functional equation and the link to Betti numbers by using the properties of étale cohomology, a new cohomology theory developed by Grothendieck and Artin for attacking the Weil conjectures, as outlined in Grothendieck (1960). Of the four conjectures the analogue of the Riemann hypothesis was the hardest to prove. Motivated by the proof of Serre (1960) of an analogue of the Weil conjectures for Kähler manifolds, Grothendieck envisioned a proof based on his standard conjectures on algebraic cycles (Kleiman 1968). However, Grothendieck's standard conjectures remain open (except for the hard Lefschetz theorem, which was proved by Deligne by extending his work on the Weil conjectures), and the analogue of the Riemann hypothesis was proved by Deligne (1974), using the étale cohomology theory but circumventing the use of standard conjectures by an ingenious argument.

Deligne (1980) found and proved a generalization of the Weil conjectures, bounding the weights of the pushforward of a sheaf.

Statement of the Weil Conjectures

Suppose that X is a non-singular n-dimensional projective algebraic variety over the field $\mathbf{F}_q$ with q elements. The **zeta function** ζ(X, s) of X is by definition

$$\zeta(X,s) = \exp\left(\sum_{m=1}^{\infty} \frac{N_m}{m} q^{-ms}\right)$$

where N_m is the number of points of X defined over the degree m extension $\mathbf{F}_{q^m}$ of $\mathbf{F}_q$.

The Weil conjectures state:

- (Rationality) ζ(X, s) is a rational function of $T = q^{-s}$. More precisely, ζ(X, s) can be written as a finite alternating product

$$\prod_{i=0}^{2n} P_i(q^{-s})^{(-1)^{i+1}} = \frac{P_1(T)\cdots P_{2n-1}(T)}{P_0(T)\cdots P_{2n}(T)},$$

- where each $P_i(T)$ is an integral polynomial. Furthermore, $P_0(T) = 1 - T$, $P_{2n}(T) = 1 - q^nT$, and for $1 \le i \le 2n - 1$, $P_i(T)$ factors over **C** as $\prod_j (1 - \alpha_{ij}T)$ for some numbers α_{ij}.
- (Functional equation and Poincaré duality) The zeta function satisfies

$$\zeta(X, n-s) = \pm q^{\frac{nE}{2} - Es}\zeta(X,s)$$

or equivalently

$$\zeta(X, q^{-n}T^{-1}) = \pm q^{\frac{nE}{2}} T^E \zeta(X,T)$$

- where E is the Euler characteristic of X. In particular, for each i, the numbers $\alpha_{2n-i,1}$, $\alpha_{2n-i,2}$, ... equal the numbers $q^n/\alpha_{i,1}$, $q^n/\alpha_{i,2}$, ... in some order.
- (Riemann hypothesis) $|\alpha_{i,j}| = q^{i/2}$ for all $1 \le i \le 2n - 1$ and all j. This implies that all zeros of $P_k(T)$ lie on the "critical line" of complex numbers s with real part k/2.
- (Betti numbers) If X is a (good) "reduction mod p" of a non-singular projective variety Y defined over a number field embedded in the field of complex numbers, then the degree of P_i is the ith Betti number of the space of complex points of Y.

Examples

The Projective Line

The simplest example (other than a point) is to take X to be the projective line. The number of points of X over a field with q^m elements is just $N_m = q^m + 1$ (where the "+ 1" comes from the "point at infinity"). The zeta function is just $1/(1 - q^{-s})(1 - q^{1-s})$. It is easy to check all parts of the Weil conjectures directly. For example, the corresponding complex variety is the Riemann sphere and its initial Betti numbers are 1, 0, 1.

Projective Space

It is not much harder to do n dimensional projective space. The number of points of X over a field with q^m elements is just $N_m = 1 + q^m + q^{2m} + ... + q^{nm}$. The zeta function is just $1/(1 - q^{-s})(1 - q^{1-s})(1 - q^{2-s}) ... (1 - q^{n-s})$.

It is again easy to check all parts of the Weil conjectures directly. (Complex projective space gives the relevant Betti numbers, which nearly determine the answer.)

The number of points on the projective line and projective space are so easy to calculate because they can be written as disjoint unions of a finite number of copies of affine spaces. It is also easy to prove the Weil conjectures for other spaces, such as Grassmannians and flag varieties, which have the same "paving" property.

Elliptic Curves

These give the first non-trivial cases of the Weil conjectures (proved by Hasse). If E is an elliptic curve over a finite field with q elements, then the number of points of E defined over the field with q^m elements is $1 - \alpha^m - \beta^m + q^m$, where α and β are complex conjugates with absolute value $\sqrt{q}$. The zeta function is

$\zeta(E,s) = (1 - \alpha q^{-s})(1 - \beta q^{-s}) / (1 - q^{-s})(1 - q^{1-s})$.

Weil Cohomology

Weil suggested that the conjectures would follow from the existence of a suitable "Weil cohomology theory" for varieties over finite fields, similar to the usual cohomology with rational coefficients for complex varieties. His idea was that if F is the Frobenius automorphism over the finite field, then the number of points of the variety X over the field of order q^m is the number of fixed points of F^m (acting on all points of the variety X defined over the algebraic closure). In algebraic topology the number of fixed points of an automorphism can be worked out using the Lefschetz fixed point theorem, given as an alternating sum of traces on the cohomology groups. So if there were similar cohomology groups for varieties over finite fields, then the zeta function could be expressed in terms of them.

The first problem with this is that the coefficient field for a Weil cohomology theory cannot be the rational numbers. To see this consider the case of a supersingular elliptic curve over a finite field of characteristic p. The endomorphism ring of this is an order in a quaternion algebra over the rationals, and should act on the first cohomology group, which should be a 2-dimensional vector space over the coefficient field by analogy with the case of a complex elliptic curve. However a quaternion algebra over the rationals cannot act on a 2-dimensional vector space over the rationals. The same argument eliminates the possibility of the coefficient field being the reals or the p-adic numbers, because the quaternion algebra is still a division algebra over these fields. However it does not eliminate the possibility that the coefficient field is the field of l-adic numbers for some prime $l \neq p$, because over these fields the division algebra splits and becomes a matrix algebra, which can act on a 2-dimensional vector space. Grothendieck and Michael Artin managed to construct suitable cohomology theories over the field of l-adic numbers for each prime l ≠ p, called l-adic cohomology.

Grothendieck's Proofs of three of the Four Conjectures

By the end of 1964 Grothendieck together with Artin and Jean-Louis Verdier proved the Weil conjectures apart from the most difficult third conjecture above (the "Riemann hypothesis" conjecture) (Grothendieck 1965). The general theorems about étale cohomology allowed Grothendieck to prove an analogue of the Lefschetz fixed point formula for the l-adic cohomology theory, and by applying it to the Frobenius automorphism F he was able to prove the conjectured formula for the zeta function:

$$\zeta(s) = \frac{P_1(T)\cdots P_{2n-1}(T)}{P_0(T)P_2(T)\cdots P_{2n}(T)}$$

where each polynomial P_i is the determinant of I – TF on the l-adic cohomology group H^i.

The rationality of the zeta function follows immediately. The functional equation for the zeta function follows from Poincaré duality for l-adic cohomology, and the relation with complex Betti numbers of a lift follows from a comparison theorem between l-adic and ordinary cohomology for complex varieties.

More generally, Grothendieck proved a similar formula for the zeta function (or "generalized L-function") of a sheaf F_0:

$$Z(X_0, F_0, t) = \prod_{x \in |X_0|} \det(1 - F_x^* t^{\deg(x)} | F_0)^{-1}$$

as a product over cohomology groups:

$$Z(X_0, F_0, t) = \prod_i \det(1 - F^* t | H_c^i(F))^{(-1)^{i+1}}$$

The special case of the constant sheaf gives the usual zeta function.

Deligne's first proof of the Riemann hypothesis conjecture

Verdier (1974), Serre (1975), Katz (1976) and Freitag & Kiehl (1988) gave expository accounts of the first proof of Deligne (1974). Much of the background in l-adic cohomology is described in (Deligne 1977).

Deligne's first proof of the remaining third Weil conjecture (the "Riemann hypothesis conjecture") used the following steps:

Use of Lefschetz Pencils

- Grothendieck expressed the zeta function in terms of the trace of Frobenius on l-adic cohomology groups, so the Weil conjectures for a d-dimensional variety V over a finite field with q elements depend on showing that the eigenvalues α of Frobenius acting on the i'th l-adic cohomology group $H^i(V)$ of V have absolute values $|\alpha|=q^{i/2}$ (for an embedding of the algebraic elements of $\mathbf{Q}_l$ into the complex numbers).

- After blowing up V and extending the base field, one may assume that the variety V has a morphism onto the projective line $\mathbf{P}^1$, with a finite number of singular fibers with very mild (quadratic) singularities. The theory of monodromy of Lefschetz pencils, introduced for complex varieties (and ordinary cohomology) by Lefschetz (1924), and extended by Grothendieck (1972) and Deligne & Katz (1973) to l-adic cohomology, relates the cohomology of V to that of its fibers. The relation depends on the space E_x of **vanishing cycles**, the subspace of the cohomology $H^{d-1}(V_x)$ of a non-singular fiber V_x, spanned by classes that vanish on singular fibers.
- The Leray spectral sequence relates the middle cohomology group of V to the cohomology of the fiber and base. The hard part to deal with is more or less a group $H^1(P^1, j_*E)$ = H1 c(U,E), where U is the points the projective line with non-singular fibers, and j is the inclusion of U into the projective line, and E is the sheaf with fibers the spaces E_x of vanishing cycles.

The Key Estimate

The heart of Deligne's proof is to show that the sheaf E over U is pure, in other words to find the absolute values of the eigenvalues of Frobenius on its stalks. This is done by studying the zeta functions of the even powers E^k of E and applying Grothendieck's formula for the zeta functions as alternating products over cohomology groups. The crucial idea of considering even k powers of E was inspired by the paper Rankin (1939), who used a similar idea with k=2 for bounding the Ramanujan tau function. Langlands (1970, section 8) pointed out that a generalization of Rankin's result for higher even values of k would imply the Ramanujan conjecture, and Deligne realized that in the case of zeta functions of varieties, Grothendieck's theory of zeta functions of sheaves provided an analogue of this generalization.

- The poles of the zeta function of E^k are found using Grothendieck's formula

$$Z(U,E^k,T)=\frac{\det(1-F^*T|H^1_c(E^k))}{\det(1-F^*T|H^0_c(E^k))\det(1-F^*T|H^2_c(E^k))}$$

and calculating the cohomology groups in the denominator explicitly. The H0 c term is usually just 1 as U is usually not compact, and the H2 c can be calculated explicitly as follows. Poincaré duality relates H2 c(E^k) to H0 (E^k), which is in turn the space of covariants of the monodromy group, which is the geometric fundamental group of U acting on the fiber of E^k at a point. The fiber of E has a bilinear form induced by cup product, which is antisymmetric if d is even, and makes E into a symplectic space. (This is a little inaccurate: Deligne did later show that $E\cap E^\perp$ = 0 by using the hard Lefschetz theorem, this requires the Weil conjectures, and the proof of the Weil conjectures really has to use a slightly more complicated argument with $E/E\cap E^\perp$ rather than E.) An argument of Kazhdan and Margulis shows that the image of the monodromy group acting on E, given by the Picard–Lefschetz formula, is Zariski dense in a symplectic group and therefore has the same invariants, which are well known from classical invariant theory. Keeping track of the action of Frobenius in this calculation shows that its eigenvalues are all $q^{k(d-1)/2+1}$, so the zeta function of $Z(E^k,T)$ has poles only at $T=1/q^{k(d-1)/2+1}$.

- The Euler product for the zeta function of E^k is

$$Z(E^k,T) = \prod_x \frac{1}{Z(E^k_x,T)}$$

If k is **even** then all the coefficients of the factors on the right (considered as power series in T) are **non-negative**; this follows by writing

$$\frac{1}{\det(1 - T^{\deg(x)} F_x|E^k)} = \exp\left(\sum_{n>0} \frac{T^n}{n} \mathrm{Trace}(F_x^n|E)^k\right)$$

and using the fact that the traces of powers of F are rational, so their k powers are non-negative as k is even. Deligne proves the rationality of the traces by relating them to numbers of points of varieties, which are always (rational) integers.

- The powers series for $Z(E^k, T)$ converges for T less than the absolute value $1/q^{k(d-1)/2+1}$ of its only possible pole. When k is even the coefficients of all its Euler factors are non-negative, so that each of the Euler factors has coefficients bounded by a constant times the coefficients of $Z(E^k, T)$ and therefore converges on the same region and has no poles in this region. So for k even the polynomials Z(Ek x, T) have no zeros in this region, or in other words the eigenvalues of Frobenius on the stalks of E^k have absolute value at most $q^{k(d-1)/2+1}$.
- This estimate can be used to find the absolute value of any eigenvalue α of Frobenius on a fiber of E as follows. For any integer k, α^k is an eigenvalue of Frobenius on a stalk of E^k, which for k even is bounded by $q^{1+k(d-1)/2}$. So

$$|\alpha^k| \le q^{k(d-1)/2+1}$$

As this is true for arbitrarily large even k, this implies that

$$|\alpha| \le q^{(d-1)/2}.$$

Poincaré duality then implies that

$$|\alpha| = q^{(d-1)/2}.$$

Completion of the Proof

The deduction of the Riemann hypothesis from this estimate is mostly a fairly straightforward use of standard techniques and is done as follows.

- The eigenvalues of Frobenius on H1 c(U,E) can now be estimated as they are the zeros of the zeta function of the sheaf E. This zeta function can be written as an Euler product of zeta functions of the stalks of E, and using the estimate for the eigenvalues on these stalks shows that this product converges for $|T|<q^{-d/2-1/2}$, so that there are no zeros of the zeta function in this region. This implies that the eigenvalues of Frobenius on E are at most $q^{d/2+1/2}$ in absolute value (in fact it will soon be seen that they have absolute value exactly $q^{d/2}$). This step of the argument is very similar to the usual proof that the Riemann zeta function has no zeros with real part greater than 1, by writing it as an Euler product.

- The conclusion of this is that the eigenvalues α of the Frobenius of a variety of even dimension d on the middle cohomology group satisfy

$$|\alpha^k| \leq q^{kd/2+1/2}$$

To obtain the Riemann hypothesis one needs to eliminate the 1/2 from the exponent. This can be done as follows. Applying this estimate to any even power V^k of V and using the Künneth formula shows that the eigenvalues of Frobenius on the middle cohomology of a variety V of any dimension d satisfy

$$|\alpha^k| \leq q^{kd/2+1/2}$$

As this is true for arbitrarily large even k, this implies that

$$|\alpha| \leq q^{d/2}$$

Poincaré duality then implies that

$$|\alpha| = q^{d/2}.$$

- This proves the Weil conjectures for the middle cohomology of a variety. The Weil conjectures for the cohomology below the middle dimension follow from this by applying the weak Lefschetz theorem, and the conjectures for cohomology above the middle dimension then follow from Poincaré duality.

Deligne's Second Proof

Deligne (1980) found and proved a generalization of the Weil conjectures, bounding the weights of the pushforward of a sheaf. In practice it is this generalization rather than the original Weil conjectures that is mostly used in applications, such as the hard Lefschetz theorem. Much of the second proof is a rearrangement of the ideas of his first proof. The main extra idea needed is an argument closely related to the theorem of Hadamard and de la Vallée Poussin, used by Deligne to show that various L-series do not have zeros with real part 1.

A constructible sheaf on a variety over a finite field is called pure of weight β if for all points x the eigenvalues of the Frobenius at x all have absolute value $N(x)^{\beta/2}$, and is called mixed of weight ≤β if it can be written as repeated extensions by pure sheaves with weights ≤β.

Deligne's theorem states that if f is a morphism of schemes of finite type over a finite field, then $R^if_!$ takes mixed sheaves of weight ≤β to mixed sheaves of weight ≤β+i.

The original Weil conjectures follow by taking f to be a morphism from a smooth projective variety to a point and considering the constant sheaf $\mathbf{Q}_l$ on the variety. This gives an upper bound on the absolute values of the eigenvalues of Frobenius, and Poincaré duality then shows that this is also a lower bound.

In general $R^i f_!$ does not take pure sheaves to pure sheaves. However it does when a suitable form of Poincaré duality holds, for example if f is smooth and proper, or if one works with perverse sheaves rather than sheaves as in Beilinson, Bernstein & Deligne (1982).

Inspired by the work of Witten (1982) on Morse theory, Laumon (1987) found another proof, using Deligne's l-adic Fourier transform, which allowed him to simplify Deligne's proof by avoiding the use of the method of Hadamard and de la Vallée Poussin. His proof generalizes the classical calculation of the absolute value of Gauss sums using the fact that the norm of a Fourier transform has a simple relation to the norm of the original function. Kiehl & Weissauer (2001) used Laumon's proof as the basis for their exposition of Deligne's theorem. Katz (2001) gave a further simplification of Laumon's proof, using monodromy in the spirit of Deligne's first proof. Kedlaya (2006) gave another proof using the Fourier transform, replacing etale cohomology with rigid cohomology.

Applications

- Deligne (1980) was able to prove the hard Lefschetz theorem (part of Grothendieck's standard conjectures) using his second proof of the Weil conjectures.
- Deligne (1971) had previously shown that the Ramanujan-Petersson conjecture follows from the Weil conjectures.
- Deligne (1974, section 8) used the Weil conjectures to prove estimates for exponential sums.

POINCARÉ CONJECTURE

In mathematics, the **Poincaré conjecture** is a theorem about the characterization of the 3-sphere, which is the hypersphere that bounds the unit ball in four-dimensional space. The conjecture states:

Every simply connected, closed 3-manifold is homeomorphic to the 3-sphere.

An equivalent form of the conjecture involves a coarser form of equivalence than homeomorphism called homotopy equivalence: if a 3-manifold is homotopy equivalent to the 3-sphere, then it is necessarily homeomorphic to it.

Originally conjectured by Henri Poincaré, the theorem concerns a space that locally looks like ordinary three-dimensional space but is connected, finite in size, and lacks any boundary (a closed 3-manifold). The Poincaré conjecture claims that if such a space has the additional property that each loop in the space can be continuously tightened to a point, then it is necessarily a three-dimensional sphere. The analogous conjectures for all higher dimensions had already been proved.

After nearly a century of effort by mathematicians, Grigori Perelman presented a proof of the conjecture in three papers made available in 2002 and 2003 on arXiv. The proof built upon the program of Richard S. Hamilton to use the Ricci flow to attempt to solve the problem. Hamilton later introduced a modification of the standard Ricci flow, called Ricci flow with surgery to systematically excise singular regions as they develop, in a controlled way, but was unable to prove this method "converged" in three dimensions. Perelman completed this portion of the proof. Several teams of mathematicians verified that Perelman's proof was correct.

The Poincaré conjecture, before being proved, was one of the most important open questions in topology. In 2000, it was named one of the seven Millennium Prize Problems, for which the Clay Mathematics Institute offered a $1,000,000 prize for the first correct solution. Perelman's work survived review and was confirmed in 2006, leading to his being offered a Fields Medal, which he declined. Perelman was awarded the Millennium Prize on March 18, 2010. On July 1, 2010, he turned down the prize saying that he believed his contribution in proving the Poincaré conjecture was no greater than Hamilton's (who first suggested using the Ricci flow for the solution). As of 2017, the Poincaré conjecture is the only solved Millennium problem.

On December 22, 2006, the journal Science honored Perelman's proof of the Poincaré conjecture as the scientific "Breakthrough of the Year", the first time this honor was bestowed in the area of mathematics.

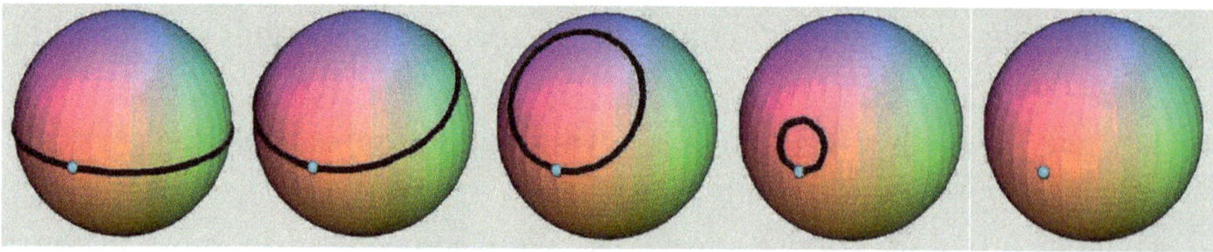

For compact 2-dimensional surfaces without boundary, if every loop can be continuously tightened to a point, then the surface is topologically homeomorphic to a 2-sphere (usually just called a sphere). The Poincaré conjecture, proved by Grigori Perelman, asserts that the same is true for 3-dimensional spaces.

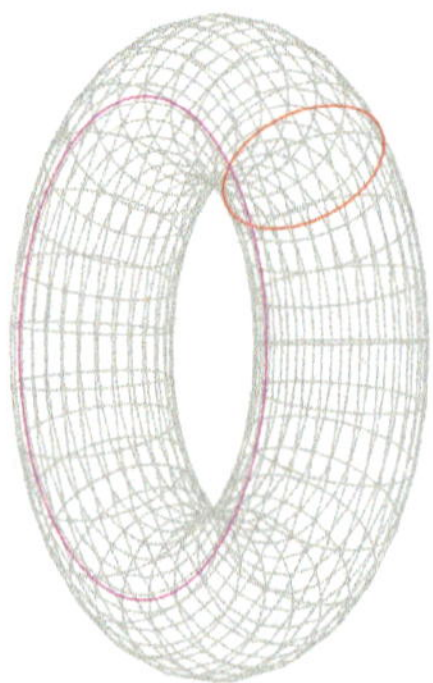

Neither of the two colored loops on this torus can be continuously tightened to a point. A torus is not homeomorphic to a sphere.

Poincaré's Question

At the beginning of the 20th century, Henri Poincaré was working on the foundations of topology—what would later be called combinatorial topology and then algebraic topology. He was particularly interested in what topological properties characterized a sphere.

Poincaré claimed in 1900 that homology, a tool he had devised based on prior work by Enrico Betti, was sufficient to tell if a 3-manifold was a 3-sphere. However, in a 1904 paper he described a counterexample to this claim, a space now called the Poincaré homology sphere. The Poincaré sphere was the first example of a homology sphere, a manifold that had the same homology as a sphere, of which many others have since been constructed. To establish that the Poincaré sphere was different from the 3-sphere, Poincaré introduced a new topological invariant, the fundamental group, and showed that the Poincaré sphere had a fundamental group of order 120, while the 3-sphere had a trivial fundamental group. In this way he was able to conclude that these two spaces were, indeed, different.

In the same paper, Poincaré wondered whether a 3-manifold with the homology of a 3-sphere and also trivial fundamental group had to be a 3-sphere. Poincaré's new condition—i.e., "trivial fundamental group"—can be restated as "every loop can be shrunk to a point."

The original phrasing was as follows:

Consider a compact 3-dimensional manifold V without boundary. Is it possible that the fundamental group of V could be trivial, even though V is not homeomorphic to the 3-dimensional sphere?

Poincaré never declared whether he believed this additional condition would characterize the 3-sphere, but nonetheless, the statement that it does is known as the Poincaré conjecture. Here is the standard form of the conjecture:

Every simply connected, closed 3-manifold is homeomorphic to the 3-sphere.

Attempted Solutions

This problem seems to have lain dormant for a time, until J. H. C. Whitehead revived interest in the conjecture, when in the 1930s he first claimed a proof, and then retracted it. In the process, he discovered some interesting examples of simply-connected non-compact 3-manifolds not homeomorphic to $\mathbf{R}^3$, the prototype of which is now called the Whitehead manifold.

In the 1950s and 1960s, other mathematicians were to claim proofs only to discover flaws in them. Influential mathematicians such as Bing, Haken, Moise, and Papakyriakopoulos attacked the conjecture. In 1958 Bing proved a weak version of the Poincaré conjecture: if every simple closed curve of a compact 3-manifold is contained in a 3-ball, then the manifold is homeomorphic to the 3-sphere. Bing also described some of the pitfalls in trying to prove the Poincaré conjecture.

Over time, the conjecture gained the reputation of being particularly tricky to tackle. John Milnor commented that sometimes the errors in false proofs can be "rather subtle and difficult to detect." Work on the conjecture improved understanding of 3-manifolds. Experts in the field were often reluctant to announce proofs, and tended to view any such announcement with skepticism. The 1980s and 1990s witnessed some well-publicized fallacious proofs (which were not actually published in peer-reviewed form).

An exposition of attempts to prove this conjecture can be found in the non-technical book Poincaré's Prize by George Szpiro.

Dimensions

The classification of closed surfaces gives an affirmative answer to the analogous question in two dimensions. For dimensions greater than three, one can pose the Generalized Poincaré conjecture: is a homotopy n-sphere homeomorphic to the n-sphere? A stronger assumption is necessary; in dimensions four and higher there are simply-connected, closed manifolds which are not homotopy equivalent to an n-sphere.

Historically, while the conjecture in dimension three seemed plausible, the generalized conjecture was thought to be false. In 1961 Stephen Smale shocked mathematicians by proving the Generalized Poincaré conjecture for dimensions greater than four and extended his techniques to prove the fundamental h-cobordism theorem. In 1982 Michael Freedman proved the Poincaré conjecture in four dimension. Freedman's work left open the possibility that there is a smooth four-manifold homeomorphic to the four-sphere which is not diffeomorphic to the four-sphere. This so-called smooth Poincaré conjecture, in dimension four, remains open and is thought to be very difficult. Milnor's exotic spheres show that the smooth Poincaré conjecture is false in dimension seven, for example.

These earlier successes in higher dimensions left the case of three dimensions in limbo. The Poincaré conjecture was essentially true in both dimension four and all higher dimensions for substantially different reasons. In dimension three, the conjecture had an uncertain reputation until the geometrization conjecture put it into a framework governing all 3-manifolds. John Morgan wrote:

It is my view that before Thurston's work on hyperbolic 3-manifolds and . . . the Geometrization conjecture there was no consensus among the experts as to whether the Poincaré conjecture was true or false. After Thurston's work, notwithstanding the fact that it had no direct bearing on the Poincaré conjecture, a consensus developed that the Poincaré conjecture (and the Geometrization conjecture) were true.

Hamilton's Program and Perelman's Solution

Hamilton's program was started in his 1982 paper in which he introduced the Ricci flow on a manifold and showed how to use it to prove some special cases of the Poincaré conjecture. In the following years he extended this work, but was unable to prove the conjecture. The actual solution was not found until Grigori Perelman published his papers.

In late 2002 and 2003 Perelman posted three papers on the arXiv. In these papers he sketched a proof of the Poincaré conjecture and a more general conjecture, Thurston's geometrization conjecture, completing the Ricci flow program outlined earlier by Richard S. Hamilton.

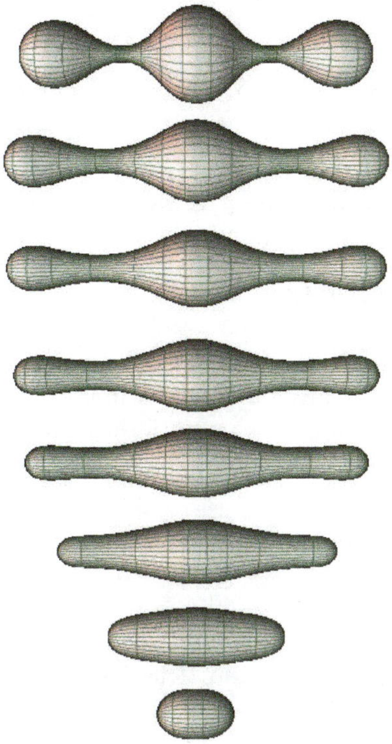

Several stages of the Ricci flow on a two-dimensional manifold

From May to July 2006, several groups presented papers that filled in the details of Perelman's proof of the Poincaré conjecture, as follows:

- Bruce Kleiner and John W. Lott posted a paper on the arXiv in May 2006 which filled in the details of Perelman's proof of the geometrization conjecture.
- Huai-Dong Cao and Xi-Ping Zhu published a paper in the June 2006 issue of the Asian Journal of Mathematics with an exposition of the complete proof of the Poincaré and geometrization conjectures.They initially implied the proof was their own achievement based on the "Hamilton-Perelman theory", but later retracted the original version of their paper, and posted a revised version, in which they referred to their work as the more modest "exposition of Hamilton–Perelman's proof". They also published an erratum disclosing that they had forgotten to cite properly the previous work of Kleiner and Lott published in 2003. In the same issue, the AJM editorial board issued an apology for what it called "incautions" in the Cao–Zhu paper.
- John Morgan and Gang Tian posted a paper on the arXiv in July 2006 which gave a detailed proof of just the Poincaré Conjecture (which is somewhat easier than the full geometrization conjecture) and expanded this to a book.

All three groups found that the gaps in Perelman's papers were minor and could be filled in using his own techniques.

On August 22, 2006, the ICM awarded Perelman the Fields Medal for his work on the conjecture, but Perelman refused the medal. John Morgan spoke at the ICM on the Poincaré conjecture on August 24, 2006, declaring that in 2003, Perelman solved the Poincaré Conjecture.

In December 2006, the journal Science honored the proof of Poincaré conjecture as the Breakthrough of the Year and featured it on its cover.

Ricci Flow with Surgery

Hamilton's program for proving the Poincaré conjecture involves first putting a Riemannian metric on the unknown simply connected closed 3-manifold. The idea is to try to improve this metric; for example, if the metric can be improved enough so that it has constant curvature, then it must be the 3-sphere. The metric is improved using the Ricci flow equations;

$$\partial_t g_{ij} = -2R_{ij}$$

where g is the metric and R its Ricci curvature, and one hopes that as the time t increases the manifold becomes easier to understand. Ricci flow expands the negative curvature part of the manifold and contracts the positive curvature part.

In some cases Hamilton was able to show that this works; for example, if the manifold has positive Ricci curvature everywhere he showed that the manifold becomes extinct in finite time under Ricci flow without any other singularities. (In other words, the manifold collapses to a point in finite time; it is easy to describe the structure just before the manifold collapses.) This easily implies the Poincaré conjecture in the case of positive Ricci curvature. However, in general the Ricci flow equations lead to singularities of the metric after a finite time. Perelman showed how to continue past these singularities: very roughly, he cuts the manifold along the singularities, splitting the manifold into several pieces, and then continues with the Ricci flow on each of these pieces. This procedure is known as Ricci flow with surgery.

A special case of Perelman's theorems about Ricci flow with surgery is given as follows.

The Ricci flow with surgery on a closed oriented 3-manifold is well defined for all time. If the fundamental group is a free product of finite groups and cyclic groups then the Ricci flow with surgery becomes extinct in finite time, and at all times all components of the manifold are connected sums of S^2 bundles over S^1 and quotients of S^3.

This result implies the Poincaré conjecture because it is easy to check it for the possible manifolds listed in the conclusion.

The condition on the fundamental group turns out to be necessary (and sufficient) for finite time extinction, and in particular includes the case of trivial fundamental group. It is equivalent to saying that the prime decomposition of the manifold has no acyclic components, and turns out to be equivalent to the condition that all geometric pieces of the manifold have geometries based on the two Thurston geometries $S^2\times\mathbf{R}$ and S^3. By studying the limit of the manifold for large time, Perelman proved Thurston's geometrization conjecture for any fundamental group: at large times the manifold has a thick-thin decomposition, whose thick piece has a hyperbolic structure, and whose thin piece is a graph manifold, but this extra complication is not necessary for proving just the Poincaré conjecture.

Solution

Grigori Perelman

In November 2002, Russian mathematician Grigori Perelman posted the first of a series of three eprints on arXiv outlining a solution of the Poincaré conjecture. Perelman's proof uses a modified version of a Ricci flow program developed by Richard S. Hamilton. In August 2006, Perelman was awarded, but declined, the Fields Medal for his proof. On March 18, 2010, the Clay Mathematics Institute awarded Perelman the $1 million Millennium Prize in recognition of his proof. Perelman rejected that prize as well.

Perelman proved the conjecture by deforming the manifold using the Ricci flow (which behaves similarly to the heat equation that describes the diffusion of heat through an object). The Ricci flow usually deforms the manifold towards a rounder shape, except for some cases where it stretches the manifold apart from itself towards what are known as singularities. Perelman and Hamilton then chop the manifold at the singularities (a process called "surgery") causing the separate pieces to form into ball-like shapes. Major steps in the proof involve showing how manifolds behave when they are deformed by the Ricci flow, examining what sort of singularities develop, determining whether this surgery process can be completed and establishing that the surgery need not be repeated infinitely many times.

The first step is to deform the manifold using the Ricci flow. The Ricci flow was defined by Richard S. Hamilton as a way to deform manifolds. The formula for the Ricci flow is an imitation of the heat equation which describes the way heat flows in a solid. Like the heat flow, Ricci flow tends towards uniform behavior. Unlike the heat flow, the Ricci flow could run into singularities and stop functioning. A singularity in a manifold is a place where it is not differentiable: like a corner or a cusp or a pinching. The Ricci flow was only defined for smooth differentiable manifolds. Hamilton used the Ricci flow to prove that some compact manifolds were diffeomorphic to spheres and he hoped to apply it to prove the Poincaré Conjecture. He needed to understand the singularities.

Hamilton created a list of possible singularities that could form but he was concerned that some singularities might lead to difficulties. He wanted to cut the manifold at the singularities and paste in caps, and then run the Ricci flow again, so he needed to understand the singularities and show that certain kinds of singularities do not occur. Perelman discovered the singularities were all very simple: essentially three-dimensional cylinders made out of spheres stretched out along a line. An ordinary cylinder is made by taking circles stretched along a line. Perelman proved this using something called the "Reduced Volume" which is closely related to an eigenvalue of a certain elliptic equation.

Sometimes an otherwise complicated operation reduces to multiplication by a scalar (a number). Such numbers are called eigenvalues of that operation. Eigenvalues are closely related to vibration frequencies and are used in analyzing a famous problem: can you hear the shape of a drum?. Essentially an eigenvalue is like a note being played by the manifold. Perelman proved this note goes up as the manifold is deformed by the Ricci flow. This helped him eliminate some of the more troublesome singularities that had concerned Hamilton, particularly the cigar soliton solution, which looked like a strand sticking out of a manifold with nothing on the other side. In essence Perelman showed that all the strands that form can be cut and capped and none stick out on one side only.

Completing the proof, Perelman takes any compact, simply connected, three-dimensional manifold without boundary and starts to run the Ricci flow. This deforms the manifold into round pieces with strands running between them. He cuts the strands and continues deforming the manifold until eventually he is left with a collection of round three-dimensional spheres. Then he rebuilds the original manifold by connecting the spheres together with three-dimensional cylinders, morphs them into a round shape and sees that, despite all the initial confusion, the manifold was in fact homeomorphic to a sphere.

One immediate question was how can one be sure there aren't infinitely many cuts necessary? Otherwise the cutting might progress forever. Perelman proved this can't happen by using minimal surfaces on the manifold. A minimal surface is essentially a soap film. Hamilton had shown that the area of a minimal surface decreases as the manifold undergoes Ricci flow. Perelman verified what happened to the area of the minimal surface when the manifold was sliced. He proved that eventually the area is so small that any cut after the area is that small can only be chopping off three-dimensional spheres and not more complicated pieces. This is described as a battle with a Hydra by Sormani in Szpiro's book cited below. This last part of the proof appeared in Perelman's third and final paper on the subject.

Shape of the Universe by Poincaré Conjecture

Friedmann-Lemaitre-Robertson-Walker universe corresponds to a time evolving radius of a S3 space. It argues that if this universe is modified in M3, at the end the acceleration it may produce a phase transition changing M3 to a space of constant curvature which corresponds precisely to a de Sitter phase associated with S3. Another point of view is that since the Geometrization conjecture (a generalization of the Poincaré conjecture) requires one to understand all locally homogeneous geometries on closed three manifolds, using Ricci flow one may consider Bianchi classification used to study cosmological models. What one may add to this scenario is that such a transition may require a torsion in order to make S3 (or other Bianchi cosmological models) parallelizable.

RIEMANN HYPOTHESIS

In mathematics, the **Riemann hypothesis** is a conjecture that the Riemann zeta function has its zeros only at the negative even integers and complex numbers with real part 1/2. It was proposed by Bernhard Riemann (1859), after whom it is named. The name is also used for some closely related analogues, such as the Riemann hypothesis for curves over finite fields.

The Riemann hypothesis implies results about the distribution of prime numbers. Along with suitable generalizations, some mathematicians consider it the most important unresolved problem in pure mathematics (Bombieri 2000). The Riemann hypothesis, along with Goldbach's conjecture, is part of Hilbert's eighth problem in David Hilbert's list of 23 unsolved problems; it is also one of the Clay Mathematics Institute's Millennium Prize Problems.

The Riemann zeta function $\zeta(s)$ is a function whose argument s may be any complex number other than 1, and whose values are also complex. It has zeros at the negative even integers; that is, $\zeta(s) = 0$ when s is one of −2, −4, −6, ... These are called its **trivial zeros**. However, the negative even integers are not the only values for which the zeta function is zero. The other ones are called non-trivial zeros. The Riemann hypothesis is concerned with the locations of these non-trivial zeros, and states that:

The real part of every non-trivial zero of the Riemann zeta function is 1/2.

Thus, if the hypothesis is correct, all the non-trivial zeros lie on the **critical line** consisting of the complex numbers $1/2 + it$, where t is a real number and i is the imaginary unit.

There are several nontechnical books on the Riemann hypothesis, such as Derbyshire (2003), Rockmore (2005), (Sabbagh 2003a, 2003b), du Sautoy (2003). The books Edwards (1974), Patterson (1988), Borwein et al. (2008) and Mazur & Stein (2015) give mathematical introductions, while Titchmarsh (1986), Ivić (1985) and Karatsuba & Voronin (1992) are advanced monographs.

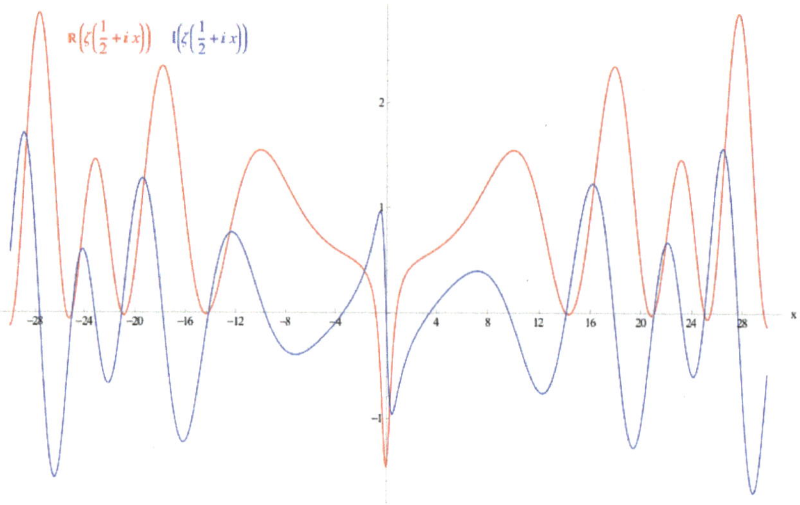

The real part (red) and imaginary part (blue) of the Riemann zeta function along the critical line Re(s) = 1/2. The first non-trivial zeros can be seen at Im(s) = ±14.135, ±21.022 and ±25.011.

Riemann Zeta Function

The Riemann zeta function is defined for complex s with real part greater than 1 by the absolutely convergent infinite series

$$\zeta(s) = \sum_{n=1}^{\infty} \frac{1}{n^s} = \frac{1}{1^s} + \frac{1}{2^s} + \frac{1}{3^s} + \cdots.$$

Leonhard Euler (who died 40 years before Riemann was born) showed that this series equals the Euler product

$$\zeta(s) = \prod_{p \text{ prime}} \frac{1}{1-p^{-s}} = \frac{1}{1-2^{-s}} \cdot \frac{1}{1-3^{-s}} \cdot \frac{1}{1-5^{-s}} \cdot \frac{1}{1-7^{-s}} \cdots \frac{1}{1-p^{-s}} \cdots$$

where the infinite product extends over all prime numbers p, and again converges for complex s with real part greater than 1. The convergence of the Euler product shows that $\zeta(s)$ has no zeros in this region, as none of the factors have zeros.

The Riemann hypothesis discusses zeros outside the region of convergence of this series and Euler product. To make sense of the hypothesis, it is necessary to analytically continue the function to give it a definition that is valid for all complex s. This can be done by expressing it in terms of the Dirichlet eta function as follows. If the real part of s is greater than one, then the zeta function satisfies

$$\left(1-\frac{2}{2^s}\right)\zeta(s)=\sum_{n=1}^{\infty}\frac{(-1)^{n+1}}{n^s}=\frac{1}{1^s}-\frac{1}{2^s}+\frac{1}{3^s}-\cdots.$$

However, the series on the right converges not just when the real part of s is greater than one, but more generally whenever s has positive real part. Thus, this alternative series extends the zeta function from Re(s) > 1 to the larger domain Re(s) > 0, excluding the zeros $s=1+2\pi in/\ln(2)$ of $1-2/2^s$ (see Dirichlet eta function). The zeta function can be extended to these values, as well, by taking limits, giving a finite value for all values of s with positive real part except for a simple pole at s = 1.

In the strip 0 < Re(s) < 1 the zeta function satisfies the functional equation

$$\zeta(s)=2^s\pi^{s-1}\sin\left(\frac{\pi s}{2}\right)\Gamma(1-s)\,\zeta(1-s).$$

One may then define ζ(s) for all remaining nonzero complex numbers s by assuming that this equation holds outside the strip as well, and letting ζ(s) equal the right-hand side of the equation whenever s has non-positive real part. If s is a negative even integer then ζ(s) = 0 because the factor sin(πs/2) vanishes; these are the **trivial zeros** of the zeta function. (If s is a positive even integer this argument does not apply because the zeros of the sine function are cancelled by the poles of the gamma function as it takes negative integer arguments.) The value ζ(0) = −1/2 is not determined by the functional equation, but is the limiting value of ζ(s) as s approaches zero. The functional equation also implies that the zeta function has no zeros with negative real part other than the trivial zeros, so all non-trivial zeros lie in the **critical strip** where s has real part between 0 and 1.

Origin

"...es ist sehr wahrscheinlich, dass alle Wurzeln reell sind. Hiervon wäre allerdings ein strenger Beweis zu wünschen; ich habe indess die Aufsuchung desselben nach einigen flüchtigen vergeblichen Versuchen vorläufig bei Seite gelassen, da er für den nächsten Zweck meiner Untersuchung entbehrlich schien." („...it is very probable that all roots are real. Of course one would wish for a rigorous proof here; I have for the time being, after some fleeting vain attempts, provisionally put aside the search for this, as it appears dispensable for the immediate objective of my investigation.")

—Riemann's statement of the Riemann hypothesis, from (Riemann 1859). (He was discussing a version of the zeta function, modified so that its roots (zeros) are real rather than on the critical line.)

Riemann's original motivation for studying the zeta function and its zeros was their occurrence in his explicit formula for the number of primes π(x) less than a given number x, which he published in his 1859 paper On the Number of Primes Less Than a Given Magnitude. His formula was given in terms of the related function

$$\Pi(x) = \pi(x) + \tfrac{1}{2}\pi(x^{\frac{1}{2}}) + \tfrac{1}{3}\pi(x^{\frac{1}{3}}) + \tfrac{1}{4}\pi(x^{\frac{1}{4}}) + \tfrac{1}{5}\pi(x^{\frac{1}{5}}) + \tfrac{1}{6}\pi(x^{\frac{1}{6}}) + \cdots$$

which counts the primes and prime powers up to x, counting a prime power p^n as 1/n of a prime. The number of primes can be recovered from this function by, using the Möbius inversion formula

$$\pi(x) = \sum_{n=1}^{\infty} \frac{\mu(n)}{n} \Pi(x^{\frac{1}{n}}) = \Pi(x) - \frac{1}{2}\Pi(x^{\frac{1}{2}}) - \frac{1}{3}\Pi(x^{\frac{1}{3}}) - \frac{1}{5}\Pi(x^{\frac{1}{5}}) + \frac{1}{6}\Pi(x^{\frac{1}{6}}) - \cdots,$$

where μ is the Möbius function. Riemann's formula is then

$$\Pi_0(x) = \mathrm{Li}(x) - \sum_{\rho} \mathrm{Li}(x^{\rho}) - \log(2) + \int_x^{\infty} \frac{dt}{t(t^2-1)\log(t)}$$

where the sum is over the nontrivial zeros of the zeta function and where Π_0 is a slightly modified version of Π that replaces its value at its points of discontinuity by the average of its upper and lower limits:

$$\Pi_0(x) = \lim_{\varepsilon \to 0} \frac{\Pi(x-\varepsilon) + \Pi(x+\varepsilon)}{2}.$$

The summation in Riemann's formula is not absolutely convergent, but may be evaluated by taking the zeros ρ in order of the absolute value of their imaginary part. The function Li occurring in the first term is the (unoffset) logarithmic integral function given by the Cauchy principal value of the divergent integral

$$\mathrm{Li}(x) = \int_0^x \frac{dt}{\log(t)}.$$

The terms $\mathrm{Li}(x^{\rho})$ involving the zeros of the zeta function need some care in their definition as Li has branch points at 0 and 1, and are defined (for $x > 1$) by analytic continuation in the complex variable ρ in the region $\mathrm{Re}(\rho) > 0$, i.e. they should be considered as $\mathrm{Ei}(\rho \ln x)$. The other terms also correspond to zeros: the dominant term Li(x) comes from the pole at $s = 1$, considered as a zero of multiplicity -1, and the remaining small terms come from the trivial zeros. For some graphs of the sums of the first few terms of this series see Riesel & Göhl (1970) or Zagier (1977).

This formula says that the zeros of the Riemann zeta function control the oscillations of primes around their expected positions. Riemann knew that the non-trivial zeros of the zeta function were symmetrically distributed about the line $s = 1/2 + it$, and he knew that all of its non-trivial zeros must lie in the range $0 \leq \mathrm{Re}(s) \leq 1$. He checked that a few of the zeros lay on the critical line with real part 1/2 and suggested that they all do; this is the Riemann hypothesis.

Consequences

The practical uses of the Riemann hypothesis include many propositions known true under the Riemann hypothesis, and some that can be shown equivalent to the Riemann hypothesis.

Distribution of Prime Numbers

Riemann's explicit formula for the number of primes less than a given number in terms of a sum over the zeros of the Riemann zeta function says that the magnitude of the oscillations of primes around their expected position is controlled by the real parts of the zeros of the zeta function. In particular the error term in the prime number theorem is closely related to the position of the zeros: for example, the supremum of real parts of the zeros is the infimum of numbers β such that the error is $O(x^\beta)$ (Ingham 1932).

Von Koch (1901) proved that the Riemann hypothesis implies the best possible bound for the error of the prime number theorem. A precise version of Koch's result, due to Schoenfeld (1976), says that the Riemann hypothesis implies

$$|\pi(x) - \mathrm{Li}(x)| < \frac{1}{8\pi}\sqrt{x}\log(x), \qquad \text{for all } x \geq 2657,$$

where $\pi(x)$ is the prime-counting function, and $\log(x)$ is the natural logarithm of x.

Schoenfeld (1976) also showed that the Riemann hypothesis implies

$$|\psi(x) - x| < \frac{1}{8\pi}\sqrt{x}\log^2(x), \qquad \text{for all } x \geq 73.2,$$

where $\psi(x)$ is Chebyshev's second function.

In 2015, Dudek proved that the Riemann hypothesis implies there is a prime p satisfying

$$x < p \leq x + \frac{4}{\pi}\sqrt{x}\log x$$

for all $x \geq 2$. This is an explicit version of a theorem of Cramer.

Growth of Arithmetic Functions

The Riemann hypothesis implies strong bounds on the growth of many other arithmetic functions, in addition to the primes counting function above.

One example involves the Möbius function μ. The statement that the equation

$$\frac{1}{\zeta(s)} = \sum_{n=1}^{\infty}\frac{\mu(n)}{n^s}$$

is valid for every s with real part greater than 1/2, with the sum on the right hand side converging, is equivalent to the Riemann hypothesis. From this we can also conclude that if the Mertens function is defined by

$$M(x) = \sum_{n \le x} \mu(n)$$

then the claim that

$$M(x) = O(x^{\frac{1}{2}+\varepsilon})$$

for every positive ε is equivalent to the Riemann hypothesis (J.E. Littlewood, 1912; see for instance: paragraph 14.25 in Titchmarsh (1986)). (For the meaning of these symbols, see Big O notation.) The determinant of the order n Redheffer matrix is equal to M(n), so the Riemann hypothesis can also be stated as a condition on the growth of these determinants. The Riemann hypothesis puts a rather tight bound on the growth of M, since Odlyzko & te Riele (1985) disproved the slightly stronger Mertens conjecture

$$|M(x)| \le \sqrt{x}.$$

The Riemann hypothesis is equivalent to many other conjectures about the rate of growth of other arithmetic functions aside from μ(n). A typical example is Robin's theorem (Robin 1984), which states that if σ(n) is the divisor function, given by

$$\sigma(n) = \sum_{d|n} d$$

then

$$\sigma(n) < e^{\gamma} n \log\log n$$

for all n > 5040 if and only if the Riemann hypothesis is true, where γ is the Euler–Mascheroni constant.

Another example was found by Jérôme Franel, and extended by Landau (see Franel & Landau (1924)). The Riemann hypothesis is equivalent to several statements showing that the terms of the Farey sequence are fairly regular. One such equivalence is as follows: if F_n is the Farey sequence of order n, beginning with 1/n and up to 1/1, then the claim that for all ε > 0

$$\sum_{i=1}^{m} |F_n(i) - \tfrac{i}{m}| = O(n^{\frac{1}{2}+\epsilon})$$

is equivalent to the Riemann hypothesis. Here

$$m = \sum_{i=1}^{n} \phi(i)$$

is the number of terms in the Farey sequence of order n.

For an example from group theory, if g(n) is Landau's function given by the maximal order of elements of the symmetric group S_n of degree n, then Massias, Nicolas & Robin (1988) showed that the Riemann hypothesis is equivalent to the bound

$$\log g(n) < \sqrt{\mathrm{Li}^{-1}(n)}$$

for all sufficiently large n.

Lindelöf Hypothesis and Growth of the Zeta Function

The Riemann hypothesis has various weaker consequences as well; one is the **Lindelöf hypothesis** on the rate of growth of the zeta function on the critical line, which says that, for any $\varepsilon > 0$,

$$\zeta\left(\frac{1}{2} + it\right) = O(t^{\varepsilon}),$$

as $t \to \infty$.

The Riemann hypothesis also implies quite sharp bounds for the growth rate of the zeta function in other regions of the critical strip. For example, it implies that

$$e^{\gamma} \leq \limsup_{t \to +\infty} \frac{|\zeta(1+it)|}{\log\log t} \leq 2e^{\gamma}$$

$$\frac{6}{\pi^2}e^{\gamma} \leq \limsup_{t \to +\infty} \frac{1/|\zeta(1+it)|}{\log\log t} \leq \frac{12}{\pi^2}e^{\gamma}$$

so the growth rate of $\zeta(1+it)$ and its inverse would be known up to a factor of 2 (Titchmarsh 1986).

Large Prime Gap Conjecture

The prime number theorem implies that on average, the gap between the prime p and its successor is log p. However, some gaps between primes may be much larger than the average. Cramér proved that, assuming the Riemann hypothesis, every gap is $O(\sqrt{p} \log p)$. This is a case in which even the best bound that can be proved using the Riemann Hypothesis is far weaker than what seems true: Cramér's conjecture implies that every gap is $O((\log p)^2)$, which, while larger than the average gap, is far smaller than the bound implied by the Riemann hypothesis. Numerical evidence supports Cramér's conjecture (Nicely 1999).

Criteria Equivalent to the Riemann Hypothesis

Many statements equivalent to the Riemann hypothesis have been found, though so far none of them have led to much progress in proving (or disproving) it. Some typical examples are as follows. (Others involve the divisor function $\sigma(n)$.)

The Riesz criterion was given by Riesz (1916), to the effect that the bound

$$-\sum_{k=1}^{\infty}\frac{(-x)^k}{(k-1)!\zeta(2k)}=O\left(x^{\frac{1}{4}+\epsilon}\right)$$

holds for all $\varepsilon > 0$ if and only if the Riemann hypothesis holds.

Nyman (1950) proved that the Riemann Hypothesis is true if and only if the space of functions of the form

$$f(x)=\sum_{\nu=1}^{n}c_\nu\rho\left(\frac{\theta_\nu}{x}\right)$$

where $\rho(z)$ is the fractional part of z, $0 \leq \theta_v \leq 1$, and

$$\sum_{\nu=1}^{n}c_\nu\theta_\nu=0,$$

is dense in the Hilbert space $L^2(0,1)$ of square-integrable functions on the unit interval. Beurling (1955) extended this by showing that the zeta function has no zeros with real part greater than 1/p if and only if this function space is dense in $L^p(0,1)$

Salem (1953) showed that the Riemann hypothesis is true if and only if the integral equation

$$\int_0^\infty\frac{z^{-\sigma-1}\phi(z)}{e^{x/z}+1}\,dz=0$$

has no non-trivial bounded solutions ϕ for $1/2<\sigma<1$.

Weil's criterion is the statement that the positivity of a certain function is equivalent to the Riemann hypothesis. Related is Li's criterion, a statement that the positivity of a certain sequence of numbers is equivalent to the Riemann hypothesis.

Speiser (1934) proved that the Riemann hypothesis is equivalent to the statement that $\zeta(s)$ the derivative of $\zeta(s)$,

$$0<\Re(s)<\frac{1}{2}.$$

That ζ has only simple zeros on the critical line is equivalent to its derivative having no zeros on the critical line.

Consequences of the Generalized Riemann Hypothesis

Several applications use the generalized Riemann hypothesis for Dirichlet L-series or zeta functions of number fields rather than just the Riemann hypothesis. Many basic properties of the Riemann zeta function can easily be generalized to all Dirichlet L-series, so it is plausible that a method that proves the Riemann hypothesis for the Riemann zeta function would also work for the generalized Riemann hypothesis for Dirichlet L-functions. Several results first proved using the generalized Riemann hypothesis were later given unconditional proofs without using it, though these were usually much harder. Many of the consequences on the following list are taken from Conrad (2010).

- In 1913, Gronwall showed that the generalized Riemann hypothesis implies that Gauss's list of imaginary quadratic fields with class number 1 is complete, though Baker, Stark and Heegner later gave unconditional proofs of this without using the generalized Riemann hypothesis.
- In 1917, Hardy and Littlewood showed that the generalized Riemann hypothesis implies a conjecture of Chebyshev that

$$\lim_{x \to 1^-} \sum_{p>2} (-1)^{(p+1)/2} x^p = +\infty,$$

which says that in some sense primes 3 mod 4 are more common than primes 1 mod 4.

- In 1923 Hardy and Littlewood showed that the generalized Riemann hypothesis implies a weak form of the Goldbach conjecture for odd numbers: that every sufficiently large odd number is the sum of three primes, though in 1937 Vinogradov gave an unconditional proof. In 1997 Deshouillers, Effinger, te Riele, and Zinoviev showed that the generalized Riemann hypothesis implies that every odd number greater than 5 is the sum of three primes.
- In 1934, Chowla showed that the generalized Riemann hypothesis implies that the first prime in the arithmetic progression a mod m is at most $Km^2\log(m)^2$ for some fixed constant K.
- In 1967, Hooley showed that the generalized Riemann hypothesis implies Artin's conjecture on primitive roots.
- In 1973, Weinberger showed that the generalized Riemann hypothesis implies that Euler's list of idoneal numbers is complete.
- Weinberger (1973) showed that the generalized Riemann hypothesis for the zeta functions of all algebraic number fields implies that any number field with class number 1 is either Euclidean or an imaginary quadratic number field of discriminant −19, −43, −67, or −163.
- In 1976, G. Miller showed that the generalized Riemann hypothesis implies that one can test if a number is prime in polynomial time via the Miller test. In 2002, Manindra Agrawal, Neeraj Kayal and Nitin Saxena proved this result unconditionally using the AKS primality test.
- Odlyzko (1990) discussed how the generalized Riemann hypothesis can be used to give sharper estimates for discriminants and class numbers of number fields.
- Ono & Soundararajan (1997) showed that the generalized Riemann hypothesis implies that Ramanujan's integral quadratic form $x^2 + y^2 + 10z^2$ represents all integers that it represents locally, with exactly 18 exceptions.

Excluded Middle

Some consequences of the RH are also consequences of its negation, and are thus theorems. In their discussion of the Hecke, Deuring, Mordell, Heilbronn theorem, (Ireland & Rosen 1990, p. 359) say

The method of proof here is truly amazing. If the generalized Riemann hypothesis is true, then the theorem is true. If the generalized Riemann hypothesis is false, then the theorem is true. Thus, the theorem is true!! (punctuation in original)

Care should be taken to understand what is meant by saying the generalized Riemann hypothesis is false: one should specify exactly which class of Dirichlet series has a counterexample.

Littlewood's Theorem

This concerns the sign of the error in the prime number theorem. It has been computed that $\pi(x) < \mathrm{Li}(x)$ for all $x \leq 10^{23}$, and no value of x is known for which $\pi(x) > \mathrm{Li}(x)$. See this table.

In 1914 Littlewood proved that there are arbitrarily large values of x for which

$$\pi(x) > \mathrm{Li}(x) + \frac{1}{3}\frac{\sqrt{x}}{\log x}\log\log\log x,$$

and that there are also arbitrarily large values of x for which

$$\pi(x) < \mathrm{Li}(x) - \frac{1}{3}\frac{\sqrt{x}}{\log x}\log\log\log x.$$

Thus the difference $\pi(x) - \mathrm{Li}(x)$ changes sign infinitely many times. Skewes' number is an estimate of the value of x corresponding to the first sign change.

Littlewood's proof is divided into two cases: the RH is assumed false (about half a page of Ingham 1932, Chapt. V), and the RH is assumed true (about a dozen pages).

Gauss's Class Number Conjecture

This is the conjecture (first stated in article 303 of Gauss's Disquisitiones Arithmeticae) that there are only a finite number of imaginary quadratic fields with a given class number. One way to prove it would be to show that as the discriminant $D \to -\infty$ the class number $h(D) \to \infty$.

The following sequence of theorems involving the Riemann hypothesis is described in Ireland & Rosen 1990, pp. 358–361:

Theorem (Hecke; 1918). Let $D < 0$ be the discriminant of an imaginary quadratic number field K. Assume the generalized Riemann hypothesis for L-functions of all imaginary quadratic Dirichlet characters. Then there is an absolute constant C such that

$$h(D) > C\frac{\sqrt{|D|}}{\log|D|}.$$

Theorem (Deuring; 1933). If the RH is false then h(D) > 1 if |D| is sufficiently large.

Theorem (Mordell; 1934). If the RH is false then h(D) → ∞ as D → −∞.

Theorem (Heilbronn; 1934). If the generalized RH is false for the L-function of some imaginary quadratic Dirichlet character then h(D) → ∞ as D → −∞.

(In the work of Hecke and Heilbronn, the only L-functions that occur are those attached to imaginary quadratic characters, and it is only for those L-functions that GRH is true or GRH is false is intended; a failure of GRH for the L-function of a cubic Dirichlet character would, strictly speaking, mean GRH is false, but that was not the kind of failure of GRH that Heilbronn had in mind, so his assumption was more restricted than simply GRH is false.)

In 1935, Carl Siegel later strengthened the result without using RH or GRH in any way.

Growth of Euler's Totient

In 1983 J. L. Nicolas proved (Ribenboim 1996, p. 320) that

$$\varphi(n) < e^{-\gamma}\frac{n}{\log\log n}$$

for infinitely many n, where φ(n) is Euler's totient function and γ is Euler's constant. Ribenboim remarks that:

The method of proof is interesting, in that the inequality is shown first under the assumption that the Riemann hypothesis is true, secondly under the contrary assumption.

Generalizations and Analogs

Dirichlet L-Series and other Number Fields

The Riemann hypothesis can be generalized by replacing the Riemann zeta function by the formally similar, but much more general, global L-functions. In this broader setting, one expects the non-trivial zeros of the global L-functions to have real part 1/2. It is these conjectures, rather than the classical Riemann hypothesis only for the single Riemann zeta function, which account for the true importance of the Riemann hypothesis in mathematics.

The generalized Riemann hypothesis extends the Riemann hypothesis to all Dirichlet L-functions. In particular it implies the conjecture that Siegel zeros (zeros of L-functions between 1/2 and 1) do not exist.

The extended Riemann hypothesis extends the Riemann hypothesis to all Dedekind zeta functions of algebraic number fields. The extended Riemann hypothesis for abelian extension of the rationals is equivalent to the generalized Riemann hypothesis. The Riemann hypothesis can also be extended to the L-functions of Hecke characters of number fields.

The grand Riemann hypothesis extends it to all automorphic zeta functions, such as Mellin transforms of Hecke eigenforms.

Function Fields and Zeta Functions of Varieties over Finite Fields

Artin (1924) introduced global zeta functions of (quadratic) function fields and conjectured an analogue of the Riemann hypothesis for them, which has been proved by Hasse in the genus 1 case and by Weil (1948) in general. For instance, the fact that the Gauss sum, of the quadratic character of a finite field of size q (with q odd), has absolute value $\sqrt{q}$ is actually an instance of the Riemann hypothesis in the function field setting. This led Weil (1949) to conjecture a similar statement for all algebraic varieties; the resulting Weil conjectures were proved by Pierre Deligne (1974, 1980).

Arithmetic Zeta Functions of Arithmetic Schemes and Their L-Factors

Arithmetic zeta functions generalise the Riemann and Dedekind zeta functions as well as the zeta functions of varieties over finite fields to every arithmetic scheme or a scheme of finite type over integers. The arithmetic zeta function of a regular connected equidimensional arithmetic scheme of Kronecker dimension n can be factorized into the product of appropriately defined L-factors and an auxiliary factor Jean-Pierre Serre (1969–1970). Assuming a functional equation and meromorphic continuation, the generalized Riemann hypothesis for the L-factor states that its zeros inside the critical strip $\Re(s) \in (0, n)$ lie on the central line. Correspondingly, the generalized Riemann hypothesis for the arithmetic zeta function of a regular connected equidimensional arithmetic scheme states that its zeros inside the critical strip lie on vertical lines $\Re(s) = 1/2, 3/2, \ldots, n - 1/2$ and its poles inside the critical strip lie on vertical lines $\Re(s) = 1, 2, \ldots, n - 1$. This is known for schemes in positive characteristic and follows from Pierre Deligne (1974, 1980), but remains entirely unknown in characteristic zero.

Selberg zeta Functions

Selberg (1956) introduced the Selberg zeta function of a Riemann surface. These are similar to the Riemann zeta function: they have a functional equation, and an infinite product similar to the Euler product but taken over closed geodesics rather than primes. The Selberg trace formula is the analogue for these functions of the explicit formulas in prime number theory. Selberg proved that the Selberg zeta functions satisfy the analogue of the Riemann hypothesis, with the imaginary parts of their zeros related to the eigenvalues of the Laplacian operator of the Riemann surface.

Ihara zeta Functions

The Ihara zeta function of a finite graph is an analogue of the Selberg zeta function, which was first introduced by Yasutaka Ihara in the context of discrete subgroups of the two-by-two p-adic special linear group. A regular finite graph is a Ramanujan graph, a mathematical model of efficient communication networks, if and only if its Ihara zeta function satisfies the analogue of the Riemann hypothesis as was pointed out by T. Sunada.

Montgomery's Pair Correlation Conjecture

Montgomery (1973) suggested the pair correlation conjecture that the correlation functions of the (suitably normalized) zeros of the zeta function should be the same as those of the eigenvalues of a random hermitian matrix. Odlyzko (1987) showed that this is supported by large scale numerical calculations of these correlation functions.

Montgomery showed that (assuming the Riemann hypothesis) at least 2/3 of all zeros are simple, and a related conjecture is that all zeros of the zeta function are simple (or more generally have no non-trivial integer linear relations between their imaginary parts). Dedekind zeta functions of algebraic number fields, which generalize the Riemann zeta function, often do have multiple complex zeros (Radziejewski 2007). This is because the Dedekind zeta functions factorize as a product of powers of Artin L-functions, so zeros of Artin L-functions sometimes give rise to multiple zeros of Dedekind zeta functions. Other examples of zeta functions with multiple zeros are the L-functions of some elliptic curves: these can have multiple zeros at the real point of their critical line; the Birch-Swinnerton-Dyer conjecture predicts that the multiplicity of this zero is the rank of the elliptic curve.

Other Zeta Functions

There are many other examples of zeta functions with analogues of the Riemann hypothesis, some of which have been proved. Goss zeta functions of function fields have a Riemann hypothesis, proved by Sheats (1998). The main conjecture of Iwasawa theory, proved by Barry Mazur and Andrew Wiles for cyclotomic fields, and Wiles for totally real fields, identifies the zeros of a p-adic L-function with the eigenvalues of an operator, so can be thought of as an analogue of the Hilbert–Pólya conjecture for p-adic L-functions (Wiles 2000).

Attempted Proofs

Several mathematicians have addressed the Riemann hypothesis, but none of their attempts have yet been accepted as correct solutions. Watkins (2007) lists some incorrect solutions, and more are frequently announced.

Operator Theory

Hilbert and Pólya suggested that one way to derive the Riemann hypothesis would be to find a self-adjoint operator, from the existence of which the statement on the real parts of the zeros of ζ(s) would follow when one applies the criterion on real eigenvalues. Some support for this idea comes from several analogues of the Riemann zeta functions whose zeros correspond to eigenvalues of some operator: the zeros of a zeta function of a variety over a finite field correspond to eigenvalues of a Frobenius element on an étale cohomology group, the zeros of a Selberg zeta function are eigenvalues of a Laplacian operator of a Riemann surface, and the zeros of a p-adic zeta function correspond to eigenvectors of a Galois action on ideal class groups.

Odlyzko (1987) showed that the distribution of the zeros of the Riemann zeta function shares some statistical properties with the eigenvalues of random matrices drawn from the Gaussian unitary ensemble. This gives some support to the Hilbert–Pólya conjecture.

In 1999, Michael Berry and Jonathan Keating conjectured that there is some unknown quantization $\hat{H}$ of the classical Hamiltonian H = xp so that $\zeta(1/2 + i\hat{H}) = 0$ and even more strongly, that the Riemann zeros coincide with the spectrum of the operator $1/2 + i\hat{H}$. This is in contrast to canonical quantization, which leads to the Heisenberg uncertainty principle $[x, p] = 1/2$ and the natural numbers as spectrum of the quantum harmonic oscillator. The crucial point is that the Hamiltonian should be a self-adjoint operator so that the quantization would be a realization of the Hilbert–Pólya program. In a connection with this quantum mechanical problem Berry and Connes had proposed that the inverse of the potential of the Hamiltonian is connected to the half-derivative of the function

$$N(s) = \frac{1}{\pi} \operatorname{Arg} \xi(1/2 + i\sqrt{s})$$

then, in Berry–Connes approach

$$V^{-1}(x) = \sqrt{4\pi} \frac{d^{1/2} N(x)}{dx^{1/2}}$$

(Connes 1999). This yields to a Hamiltonian whose eigenvalues are the square of the imaginary part of the Riemann zeros, and also the functional determinant of this Hamiltonian operator is just the Riemann Xi function. In fact the Riemann Xi function would be proportional to the functional determinant (Hadamard product)

$$\det(H + 1/4 + s(s-1))$$

as proved by Connes and others, in this approach

$$\frac{\xi(s)}{\xi(0)} = \frac{\det(H + s(s-1) + 1/4)}{\det(H + 1/4)}.$$

The analogy with the Riemann hypothesis over finite fields suggests that the Hilbert space containing eigenvectors corresponding to the zeros might be some sort of first cohomology group of the spectrum Spec(**Z**) of the integers. Deninger (1998) described some of the attempts to find such a cohomology theory (Leichtnam 2005).

Zagier (1981) constructed a natural space of invariant functions on the upper half plane that has eigenvalues under the Laplacian operator that correspond to zeros of the Riemann zeta function—and remarked that in the unlikely event that one could show the existence of a suitable positive definite inner product on this space, the Riemann hypothesis would follow. Cartier (1982) discussed a related example, where due to a bizarre bug a computer program listed zeros of the Riemann zeta function as eigenvalues of the same Laplacian operator.

Schumayer & Hutchinson (2011) surveyed some of the attempts to construct a suitable physical model related to the Riemann zeta function.

Lee–Yang Theorem

The Lee–Yang theorem states that the zeros of certain partition functions in statistical mechanics all lie on a "critical line" with real part 0, and this has led to some speculation about a relationship with the Riemann hypothesis (Knauf 1999).

Turán's Result

Pál Turán (1948) showed that if the functions

$$\sum_{n=1}^{N} n^{-s}$$

have no zeros when the real part of s is greater than one then

$$T(x) = \sum_{n \leq x} \frac{\lambda(n)}{n} \geq 0 \text{ for } x > 0,$$

where $\lambda(n)$ is the Liouville function given by $(-1)^r$ if n has r prime factors. He showed that this in turn would imply that the Riemann hypothesis is true. However Haselgrove (1958) proved that T(x) is negative for infinitely many x (and also disproved the closely related Pólya conjecture), and Borwein, Ferguson & Mossinghoff (2008) showed that the smallest such x is 72185376951205. Spira (1968) showed by numerical calculation that the finite Dirichlet series above for N=19 has a zero with real part greater than 1. Turán also showed that a somewhat weaker assumption, the nonexistence of zeros with real part greater than $1+N^{-1/2+\varepsilon}$ for large N in the finite Dirichlet series above, would also imply the Riemann hypothesis, but Montgomery (1983) showed that for all sufficiently large N these series have zeros with real part greater than 1 + (log log N)/(4 log N). Therefore, Turán's result is vacuously true and cannot be used to help prove the Riemann hypothesis.

Noncommutative Geometry

Connes (1999, 2000) has described a relationship between the Riemann hypothesis and noncommutative geometry, and shows that a suitable analog of the Selberg trace formula for the action of the idèle class group on the adèle class space would imply the Riemann hypothesis. Some of these ideas are elaborated in Lapidus (2008).

Hilbert Spaces of Entire Functions

Louis de Branges (1992) showed that the Riemann hypothesis would follow from a positivity condition on a certain Hilbert space of entire functions. However Conrey & Li (2000) showed that the necessary positivity conditions are not satisfied. Despite this obstacle, de Branges has continued to work on an attempted proof of the Riemann hypothesis along the same lines, but this has not been widely accepted by other mathematicians (Sarnak 2005).

Quasicrystals

The Riemann hypothesis implies that the zeros of the zeta function form a quasicrystal, meaning a distribution with discrete support whose Fourier transform also has discrete support. Dyson (2009) suggested trying to prove the Riemann hypothesis by classifying, or at least studying, 1-dimensional quasicrystals.

Arithmetic Zeta Functions of Models of Elliptic Curves Over Number Fields

When one goes from geometric dimension one, e.g. an algebraic number field, to geometric dimension two, e.g. a regular model of an elliptic curve over a number field, the two-dimensional part of the generalized Riemann hypothesis for the arithmetic zeta function of the model deals with the poles of the zeta function. In dimension one the study of the zeta integral in Tate's thesis does not lead to new important information on the Riemann hypothesis. Contrary to this, in dimension two work of Ivan Fesenko on two-dimensional generalisation of Tate's thesis includes an integral representation of a zeta integral closely related to the zeta function. In this new situation, not possible in dimension one, the poles of the zeta function can be studied via the zeta integral and associated adele groups. Related conjecture of Fesenko (2010) on the positivity of the fourth derivative of a boundary function associated to the zeta integral essentially implies the pole part of the generalized Riemann hypothesis. Suzuki (2011) proved that the latter, together with some technical assumptions, implies Fesenko's conjecture.

Multiple Zeta Functions

Deligne's proof of the Riemann hypothesis over finite fields used the zeta functions of product varieties, whose zeros and poles correspond to sums of zeros and poles of the original zeta function, in order to bound the real parts of the zeros of the original zeta function. By analogy, Kurokawa (1992) introduced multiple zeta functions whose zeros and poles correspond to sums of zeros and poles of the Riemann zeta function. To make the series converge he restricted to sums of zeros or poles all with non-negative imaginary part. So far, the known bounds on the zeros and poles of the multiple zeta functions are not strong enough to give useful estimates for the zeros of the Riemann zeta function.

Location of the Zeros

Number of Zeros

The functional equation combined with the argument principle implies that the number of zeros of the zeta function with imaginary part between 0 and T is given by

$$N(T) = \frac{1}{\pi}\mathrm{Arg}(\xi(s)) = \frac{1}{\pi}\mathrm{Arg}(\Gamma(\tfrac{s}{2})\pi^{-\frac{s}{2}}\zeta(s)s(s-1)/2)$$

for s=1/2+iT, where the argument is defined by varying it continuously along the line with Im(s)=T, starting with argument 0 at ∞+iT. This is the sum of a large but well understood term

$$\frac{1}{\pi}\mathrm{Arg}(\Gamma(\tfrac{s}{2})\pi^{-s/2}s(s-1)/2) = \frac{T}{2\pi}\log\frac{T}{2\pi} - \frac{T}{2\pi} + 7/8 + O(1/T)$$

and a small but rather mysterious term

$$S(T) = \frac{1}{\pi}\mathrm{Arg}(\zeta(1/2+iT)) = O(\log(T)).$$

So the density of zeros with imaginary part near T is about log(T)/2π, and the function S describes the small deviations from this. The function S(t) jumps by 1 at each zero of the zeta function, and for $t \geq 8$ it decreases monotonically between zeros with derivative close to −log t.

Karatsuba (1996) proved that every interval $(T, T+H]$ for $H \geq T^{\frac{27}{82}+\varepsilon}$ contains at least

$$H(\ln T)^{\frac{1}{3}}e^{-c\sqrt{\ln\ln T}}$$

points where the function S(t) changes sign.

Selberg (1946) showed that the average moments of even powers of S are given by

$$\int_0^T |S(t)|^{2k}dt = \frac{(2k)!}{k!(2\pi)^{2k}}T(\log\log T)^k + O(T(\log\log T)^{k-1/2}).$$

This suggests that $S(T)/(\log\log T)^{1/2}$ resembles a Gaussian random variable with mean 0 and variance $2\pi^2$ (Ghosh (1983) proved this fact). In particular |S(T)| is usually somewhere around $(\log\log T)^{1/2}$, but occasionally much larger. The exact order of growth of S(T) is not known. There has been no unconditional improvement to Riemann's original bound S(T)=O(log T), though the Riemann hypothesis implies the slightly smaller bound S(T)=O(log T/log log T) (Titchmarsh 1986). The true order of magnitude may be somewhat less than this, as random functions with the same distribution as S(T) tend to have growth of order about $\log(T)^{1/2}$. In the other direction it cannot be too small: Selberg (1946) showed that $S(T) \neq o((\log T)^{1/3}/(\log\log T)^{7/3})$, and assuming the Riemann hypothesis Montgomery showed that $S(T) \neq o((\log T)^{1/2}/(\log\log T)^{1/2})$.

Numerical calculations confirm that S grows very slowly: $|S(T)| < 1$ for $T < 280$, $|S(T)| < 2$ for $T < 6800000$, and the largest value of $|S(T)|$ found so far is not much larger than 3 (Odlyzko 2002).

Riemann's estimate S(T) = O(log T) implies that the gaps between zeros are bounded, and Littlewood improved this slightly, showing that the gaps between their imaginary parts tends to 0.

Theorem of Hadamard and De La Vallée-Poussin

Hadamard (1896) and de la Vallée-Poussin (1896) independently proved that no zeros could lie on the line Re(s) = 1. Together with the functional equation and the fact that there are no zeros with real part greater than 1, this showed that all non-trivial zeros must lie in the interior of the critical strip 0 < Re(s) < 1. This was a key step in their first proofs of the prime number theorem.

Both the original proofs that the zeta function has no zeros with real part 1 are similar, and depend on showing that if ζ(1+it) vanishes, then ζ(1+2it) is singular, which is not possible. One way of doing this is by using the inequality

$$|\zeta(\sigma)^3\zeta(\sigma+it)^4\zeta(\sigma+2it)| \geq 1$$

for $\sigma > 1$, t real, and looking at the limit as $\sigma \to 1$. This inequality follows by taking the real part of the log of the Euler product to see that

$$|\zeta(\sigma+it)| = \exp\Re\sum_{p^n}\frac{p^{-n(\sigma+it)}}{n} = \exp\sum_{p^n}\frac{p^{-n\sigma}\cos(t\log p^n)}{n},$$

where the sum is over all prime powers p^n, so that

$$|\zeta(\sigma)^3\zeta(\sigma+it)^4\zeta(\sigma+2it)| = \exp\sum_{p^n}p^{-n\sigma}\frac{3+4\cos(t\log p^n)+\cos(2t\log p^n)}{n}$$

which is at least 1 because all the terms in the sum are positive, due to the inequality

$$3+4\cos(\theta)+\cos(2\theta) = 2(1+\cos(\theta))^2 \geq 0.$$

Zero-Free Regions

De la Vallée-Poussin (1899–1900) proved that if $\sigma + it$ is a zero of the Riemann zeta function, then $1 - \sigma \geq C/\log(t)$ for some positive constant C. In other words, zeros cannot be too close to the line $\sigma = 1$: there is a zero-free region close to this line. This zero-free region has been enlarged by several authors using methods such as Vinogradov's mean-value theorem. Ford (2002) gave a version with explicit numerical constants: $\zeta(\sigma + it) \neq 0$ whenever $|t| \geq 3$ and

$$\sigma \geq 1 - \frac{1}{57.54(\log|t|)^{2/3}(\log\log|t|)^{1/3}}.$$

Zeros on the Critical Line

Hardy (1914) and Hardy & Littlewood (1921) showed there are infinitely many zeros on the critical line, by considering moments of certain functions related to the zeta function. Selberg (1942) proved that at least a (small) positive proportion of zeros lie on the line. Levinson (1974) improved this to one-third of the zeros by relating the zeros of the zeta function to those of its derivative, and Conrey (1989) improved this further to two-fifths.

Most zeros lie close to the critical line. More precisely, Bohr & Landau (1914) showed that for any positive ε, all but an infinitely small proportion of zeros lie within a distance ε of the critical line. Ivić (1985) gives several more precise versions of this result, called zero density estimates, which bound the number of zeros in regions with imaginary part at most T and real part at least 1/2+ε.

Hardy–Littlewood Conjectures

In 1914 Godfrey Harold Hardy proved that $\zeta\left(\frac{1}{2}+it\right)$ has infinitely many real zeros. Let N(T) be the total number of real zeros, $N_0(T)$ be the total number of zeros of odd order of the function

$$\zeta\left(\frac{1}{2}+it\right),$$

lying on the interval (0, T].

The next two conjectures of Hardy and John Edensor Littlewood on the distance between real zeros of $\zeta\left(\frac{1}{2}+it\right)$ and on the density of zeros of $\zeta\left(\frac{1}{2}+it\right)$ on intervals (T, T+H] for sufficiently large T > 0, $H = T^{a+\varepsilon}$ and with as less as possible value of a > 0, where ε > 0 is an arbitrarily small number, open two new directions in the investigation of the Riemann zeta function:

1. for any ε > 0 there exists $T_0 = T_0(\varepsilon) > 0$ such that for $T \geq T_0$ and $H = T^{0.25+\varepsilon}$ the interval $(T, T+H]$ contains a zero of odd order of the function $\zeta\left(\frac{1}{2}+it\right)$.

2. for any ε > 0 there exist $T_0 = T_0(\varepsilon) > 0$ and c = c(ε) > 0, such that for $T \geq T_0$ and $H = T^{0.5+\varepsilon}$ the inequality $N_0(T+H) - N_0(T) \geq cH$ is true.

Selberg's Zeta Function Conjecture

Atle Selberg (1942) investigated the problem of Hardy–Littlewood **2** and proved that for any ε > 0 there exists such $T_0 = T_0(\varepsilon) > 0$ and c = c(ε) > 0, such that for $T \geq T_0$ and $H = T^{0.5+\varepsilon}$ the inequality $N(T+H) - N(T) \geq cH\log T$ is true. Selberg conjectured that this could be tightened to $H = T^{0.5}$. A. A. Karatsuba (1984a, 1984b, 1985) proved that for a fixed ε satisfying the condition 0 < ε < 0.001, a sufficiently large T and $H = T^{a+\varepsilon}$, $a = \frac{27}{82} = \frac{1}{3} - \frac{1}{246}$, the interval (T, T+H) contains at least cHln(T) real zeros of the Riemann zeta function $\zeta\left(\frac{1}{2}+it\right)$ and therefore confirmed the Selberg conjecture. The estimates of Selberg and Karatsuba can not be improved in respect of the order of growth as T → ∞.

Karatsuba (1992) proved that an analog of the Selberg conjecture holds for almost all intervals (T, T+H], $H = T^{\varepsilon}$ where ε is an arbitrarily small fixed positive number. The Karatsuba method permits to investigate zeros of the Riemann zeta-function on "supershort" intervals of the critical line, that is, on the intervals (T, T+H], the length H of which grows slower than any, even arbitrarily small degree T. In particular, he proved that for any given numbers ε, ε_1 satisfying the conditions $0 < \varepsilon, \varepsilon_1 < 1$ almost all intervals (T, T+H] for $(T, T+H]$ for $H \geq \exp\{(\ln T)^{\varepsilon}\}$ contain at least $H(\ln T)^{1-\varepsilon_1}$ zeros of the function $\zeta\left(\frac{1}{2}+it\right)$ This estimate is quite close to the one that follows from the Riemann hypothesis.

Numerical Calculations

The function

$$\pi^{-\frac{s}{2}}\Gamma(\tfrac{s}{2})\zeta(s)$$

has the same zeros as the zeta function in the critical strip, and is real on the critical line because of the functional equation, so one can prove the existence of zeros exactly on the real line between two points by checking numerically that the function has opposite signs at these points. Usually one writes

$$\zeta(\tfrac{1}{2}+it) = Z(t)e^{-i\theta(t)}$$

where Hardy's function Z and the Riemann–Siegel theta function θ are uniquely defined by this and the condition that they are smooth real functions with θ(0)=0.

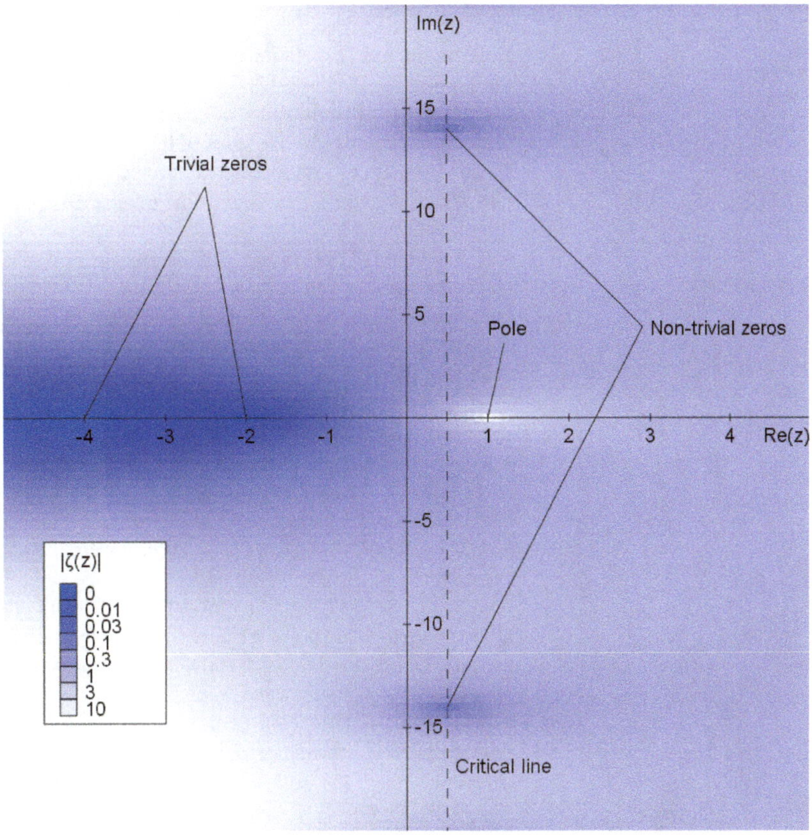

Absolute value of the ζ-function

By finding many intervals where the function Z changes sign one can show that there are many zeros on the critical line. To verify the Riemann hypothesis up to a given imaginary part T of the zeros, one also has to check that there are no further zeros off the line in this region. This can be done by calculating the total number of zeros in the region and checking that it is the same as the number of zeros found on the line. This allows one to verify the Riemann hypothesis computationally up to any desired value of T (provided all the zeros of the zeta function in this region are simple and on the critical line).

Some calculations of zeros of the zeta function are listed below. So far all zeros that have been checked are on the critical line and are simple. (A multiple zero would cause problems for the zero finding algorithms, which depend on finding sign changes between zeros.) For tables of the zeros, see Haselgrove & Miller (1960) or Odlyzko.

Year	Number of zeros	Author
1859?	3	B. Riemann used the Riemann–Siegel formula (unpublished, but reported in Siegel 1932).
1903	15	J. P. Gram (1903) used Euler–Maclaurin summation and discovered Gram's law. He showed that all 10 zeros with imaginary part at most 50 range lie on the critical line with real part 1/2 by computing the sum of the inverse 10th powers of the roots he found.
1914	79 ($\gamma_n \leq 200$)	R. J. Backlund (1914) introduced a better method of checking all the zeros up to that point are on the line, by studying the argument S(T) of the zeta function.
1925	138 ($\gamma_n \leq 300$)	J. I. Hutchinson (1925) found the first failure of Gram's law, at the Gram point g_{126}.
1935	195	E. C. Titchmarsh (1935) used the recently rediscovered Riemann–Siegel formula, which is much faster than Euler–Maclaurin summation. It takes about $O(T^{3/2+\varepsilon})$ steps to check zeros with imaginary part less than T, while the Euler–Maclaurin method takes about $O(T^{2+\varepsilon})$ steps.
1936	1041	E. C. Titchmarsh (1936) and L. J. Comrie were the last to find zeros by hand.
1953	1104	A. M. Turing (1953) found a more efficient way to check that all zeros up to some point are accounted for by the zeros on the line, by checking that Z has the correct sign at several consecutive Gram points and using the fact that S(T) has average value 0. This requires almost no extra work because the sign of Z at Gram points is already known from finding the zeros, and is still the usual method used. This was the first use of a digital computer to calculate the zeros.
1956	15000	D. H. Lehmer (1956) discovered a few cases where the zeta function has zeros that are "only just" on the line: two zeros of the zeta function are so close together that it is unusually difficult to find a sign change between them. This is called "Lehmer's phenomenon", and first occurs at the zeros with imaginary parts 7005.063 and 7005.101, which differ by only .04 while the average gap between other zeros near this point is about 1.
1956	25000	D. H. Lehmer
1958	35337	N. A. Meller
1966	250000	R. S. Lehman
1968	3500000	Rosser, Yohe & Schoenfeld (1969) stated Rosser's rule (described below).
1977	40000000	R. P. Brent
1979	81000001	R. P. Brent
1982	200000001	R. P. Brent, J. van de Lune, H. J. J. te Riele, D. T. Winter
1983	300000001	J. van de Lune, H. J. J. te Riele
1986	1500000001	van de Lune, te Riele & Winter (1986) gave some statistical data about the zeros and give several graphs of Z at places where it has unusual behavior.
1987	A few of large ($\sim 10^{12}$) height	A. M. Odlyzko (1987) computed smaller numbers of zeros of much larger height, around 10^{12}, to high precision to check Montgomery's pair correlation conjecture.
1992	A few of large ($\sim 10^{20}$) height	A. M. Odlyzko (1992) computed a 175 million zeroes of heights around 10^{20} and a few more of heights around 2×10^{20}, and gave an extensive discussion of the results.

1998	10000 of large (~10^{21}) height	A. M. Odlyzko (1998) computed some zeros of height about 10^{21}
2001	10000000000	J. van de Lune (unpublished)
2004	900000000000	S. Wedeniwski (ZetaGrid distributed computing)
2004	10000000000000and a few of large (up to ~10^{24}) heights	X. Gourdon (2004) and Patrick Demichel used the Odlyzko–Schönhage algorithm. They also checked two billion zeros around heights 10^{13}, 10^{14}, ..., 10^{24}.

Gram Points

A Gram point is a point on the critical line 1/2 + it where the zeta function is real and non-zero. Using the expression for the zeta function on the critical line, $\zeta(1/2 + it) = Z(t)\, e^{-i\theta(t)}$, where Hardy's function, Z, is real for real t, and θ is the Riemann–Siegel theta function, we see that zeta is real when $\sin(\theta(t)) = 0$. This implies that θ(t) is an integer multiple of π, which allows for the location of Gram points to be calculated fairly easily by inverting the formula for θ. They are usually numbered as g_n for n = 0, 1, ..., where g_n is the unique solution of $\theta(t) = n\pi$.

Gram observed that there was often exactly one zero of the zeta function between any two Gram points; Hutchinson called this observation **Gram's law**. There are several other closely related statements that are also sometimes called Gram's law: for example, $(-1)^n Z(g_n)$ is usually positive, or Z(t) usually has opposite sign at consecutive Gram points. The imaginary parts γ_n of the first few zeros (in blue) and the first few Gram points g_n are given in the following table

		g_{-1}	γ_1	g_0	γ_2	g_1	γ_3	g_2	γ_4	g_3	γ_5	g_4	γ_6	g_5
0.000	3.436	9.667	14.135	17.846	21.022	23.170	25.011	27.670	30.425	31.718	32.935	35.467	37.586	38.999

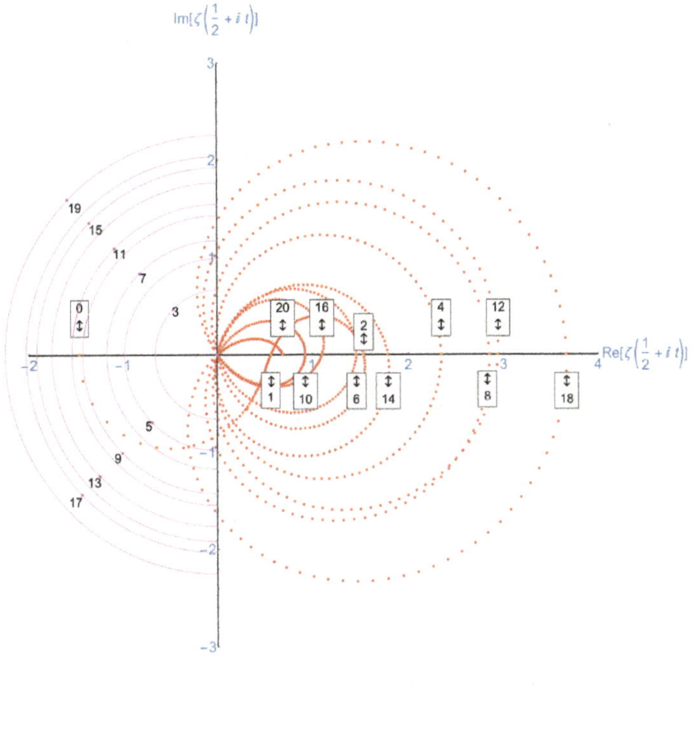

This is a polar plot of the first 20 non-trivial Riemann zeta function zeros (including Gram points) along the critical line $\zeta(1/2 + it)$ for real values of t running from 0 to 50. The consecutively labeled zeros have 50 red plot points between each, with zeros identified by concentric magenta rings scaled to show the relative distance between their values of t. Gram's law states that the curve usually crosses the real axis once between zeros.

The first failure of Gram's law occurs at the 127'th zero and the Gram point g_{126}, which are in the "wrong" order.

g_{124}	γ_{126}	g_{125}	g_{126}	γ_{127}	γ_{128}	g_{127}	γ_{129}	g_{128}
279.148	279.229	280.802	282.455	**282.465**	283.211	284.104	284.836	285.752

A Gram point t is called good if the zeta function is positive at 1/2 + it. The indices of the "bad" Gram points where Z has the "wrong" sign are 126, 134, 195, 211, ... (sequence A114856 in the OEIS). A **Gram block** is an interval bounded by two good Gram points such that all the Gram points between them are bad. A refinement of Gram's law called Rosser's rule due to Rosser, Yohe & Schoenfeld (1969) says that Gram blocks often have the expected number of zeros in them (the same as the number of Gram intervals), even though some of the individual Gram intervals in the block may not have exactly one zero in them. For example, the interval bounded by g_{125} and g_{127} is a Gram block containing a unique bad Gram point g_{126}, and contains the expected number 2 of zeros although neither of its two Gram intervals contains a unique zero. Rosser et al. checked that there were no exceptions to Rosser's rule in the first 3 million zeros, although there are infinitely many exceptions to Rosser's rule over the entire zeta function.

Gram's rule and Rosser's rule both say that in some sense zeros do not stray too far from their expected positions. The distance of a zero from its expected position is controlled by the function S defined above, which grows extremely slowly: its average value is of the order of $(\log \log T)^{1/2}$, which only reaches 2 for T around 10^{24}. This means that both rules hold most of the time for small T but eventually break down often. Indeed, Trudgian (2011) showed that both Gram's law and Rosser's rule fail in a positive proportion of cases. To be specific, it is expected that in about 73% one zero is enclosed by two successive Gram points, but in 14% no zero and in 13% two zeros are in such a Gram-interval on the long run.

Arguments for and Against the Riemann Hypothesis

Mathematical papers about the Riemann hypothesis tend to be cautiously noncommittal about its truth. Of authors who express an opinion, most of them, such as Riemann (1859) or Bombieri (2000), imply that they expect (or at least hope) that it is true. The few authors who express serious doubt about it include Ivić (2008) who lists some reasons for being skeptical, and Littlewood (1962) who flatly states that he believes it is false, and that there is no evidence whatever for it and no imaginable reason that it would be true. The consensus of the survey articles (Bombieri 2000, Conrey 2003, and Sarnak 2005) is that the evidence for it is strong but not overwhelming, so that while it is probably true there is some reasonable doubt about it.

Some of the arguments for (or against) the Riemann hypothesis are listed by Sarnak (2005), Conrey (2003), and Ivić (2008), and include the following reasons.

- Several analogues of the Riemann hypothesis have already been proved. The proof of the Riemann hypothesis for varieties over finite fields by Deligne (1974) is possibly the single strongest theoretical reason in favor of the Riemann hypothesis. This provides some evidence for the more general conjecture that all zeta functions associated with automorphic forms satisfy a Riemann hypothesis, which includes the classical Riemann hypothesis as a special case. Similarly Selberg zeta functions satisfy the analogue of the Riemann hypothesis, and are in some ways similar to the Riemann zeta function, having a functional equation and an infinite product expansion analogous to the Euler product expansion. However, there are also some major differences; for example they are not given by Dirichlet series. The Riemann hypothesis for the Goss zeta function was proved by Sheats (1998). In contrast to these positive examples, however, some Epstein zeta functions do not satisfy the Riemann hypothesis, even though they have an infinite number of zeros on the critical line (Titchmarsh 1986). These functions are quite similar to the Riemann zeta function, and have a Dirichlet series expansion and a functional equation, but the ones known to fail the Riemann hypothesis do not have an Euler product and are not directly related to automorphic representations.
- At first, the numerical verification that many zeros lie on the line seems strong evidence for it. However, analytic number theory has had many conjectures supported by large amounts of numerical evidence that turn out to be false. See Skewes number for a notorious example, where the first exception to a plausible conjecture related to the Riemann hypothesis probably occurs around 10^{316}; a counterexample to the Riemann hypothesis with imaginary part this size would be far beyond anything that can currently be computed using a direct approach. The problem is that the behavior is often influenced by very slowly increasing functions such as log log T, that tend to infinity, but do so so slowly that this cannot be detected by computation. Such functions occur in the theory of the zeta function controlling the behavior of its zeros; for example the function S(T) above has average size around $(\log \log T)^{1/2}$. As S(T) jumps by at least 2 at any counterexample to the Riemann hypothesis, one might expect any counterexamples to the Riemann hypothesis to start appearing only when S(T) becomes large. It is never much more than 3 as far as it has been calculated, but is known to be unbounded, suggesting that calculations may not have yet reached the region of typical behavior of the zeta function.
- Denjoy's probabilistic argument for the Riemann hypothesis (Edwards 1974) is based on the observation that if μ(x) is a random sequence of "1"s and "−1"s then, for every ε > 0, the partial sums

$$M(x) = \sum_{n \le x} \mu(n)$$

(the values of which are positions in a simple random walk) satisfy the bound

$$M(x) = O(x^{1/2+\varepsilon})$$

with probability 1. The Riemann hypothesis is equivalent to this bound for the Möbius function μ and the Mertens function M derived in the same way from it. In other words, the Riemann hypothesis is in some sense equivalent to saying that μ(x) behaves like a random sequence of coin tosses. When μ(x) is non-zero its sign gives the parity of the number of prime factors of x, so informally the Riemann hypothesis says that the parity of the number of prime factors of an integer behaves randomly. Such probabilistic arguments in number theory often give the right answer, but tend to be very hard to make rigorous, and occasionally give the wrong answer for some results, such as Maier's theorem.

- The calculations in Odlyzko (1987) show that the zeros of the zeta function behave very much like the eigenvalues of a random Hermitian matrix, suggesting that they are the eigenvalues of some self-adjoint operator, which would imply the Riemann hypothesis. However all attempts to find such an operator have failed.
- There are several theorems, such as the weak Goldbach conjecture for sufficiently large odd numbers, that were first proved using the generalized Riemann hypothesis, and later shown to be true unconditionally. This could be considered as weak evidence for the generalized Riemann hypothesis, as several of its "predictions" turned out to be true.
- Lehmer's phenomenon (Lehmer 1956) where two zeros are sometimes very close is sometimes given as a reason to disbelieve in the Riemann hypothesis. However one would expect this to happen occasionally just by chance even if the Riemann hypothesis were true, and Odlyzko's calculations suggest that nearby pairs of zeros occur just as often as predicted by Montgomery's conjecture.
- Patterson (1988) suggests that the most compelling reason for the Riemann hypothesis for most mathematicians is the hope that primes are distributed as regularly as possible.

Other Conjectures

- Goldbach's conjecture
- The twin prime conjecture
- The Collatz conjecture
- The Manin conjecture
- The Maldacena conjecture
- The Langlands program is a far-reaching web of these ideas of 'unifying conjectures' that link different subfields of mathematics, e.g. number theory and representation theory of Lie groups; some of these conjectures have since been proved.

Twin Prime

A **twin prime** is a prime number that is either 2 less or 2 more than another prime number—for example, the twin prime pair (41, 43). In other words, a twin prime is a prime that has a prime gap of two. Sometimes the term twin prime is used for a pair of twin primes; an alternative name for this is **prime twin** or **prime pair**.

Twin primes become increasingly rare as one examines larger ranges, in keeping with the general tendency of gaps between adjacent primes to become larger as the numbers themselves get larger. However, it is a longstanding conjecture that there are infinitely many twin primes. Work of Yitang Zhang in 2013, as well as work by James Maynard, Terence Tao and others, has made substantial progress towards proving this conjecture, but at present it remains unsolved.

The question of whether there exist infinitely many twin primes has been one of the great open questions in number theory for many years. This is the content of the **twin prime conjecture**, which states that there are infinitely many primes p such that p + 2 is also prime. In 1849, de Polignac made the more general conjecture that for every natural number k, there are infinitely many primes p such that p + 2k is also prime. The case k = 1 is the twin prime conjecture.

A stronger form of the twin prime conjecture, the Hardy–Littlewood conjecture (see below), postulates a distribution law for twin primes akin to the prime number theorem.

On April 17, 2013, Yitang Zhang announced a proof that for some integer N that is less than 70 million, there are infinitely many pairs of primes that differ by N. Zhang's paper was accepted by Annals of Mathematics in early May 2013. Terence Tao subsequently proposed a Polymath Project collaborative effort to optimize Zhang's bound. As of April 14, 2014, one year after Zhang's announcement, the bound has been reduced to 246. Further, assuming the Elliott–Halberstam conjecture and its generalized form, the Polymath project wiki states that the bound has been reduced to 12 and 6, respectively. These improved bounds were discovered using a different approach that was simpler than Zhang's and was discovered independently by James Maynard and Terence Tao. This second approach also gave bounds for the smallest f(m) needed to guarantee that infinitely many intervals of width f(m) contain at least m primes.

Properties

Usually the pair (2, 3) is not considered to be a pair of twin primes. Since 2 is the only even prime, this pair is the only pair of prime numbers that differ by one; thus twin primes are as closely spaced as possible for two primes.

The first few twin prime pairs are:

(3, 5), (5, 7), (11, 13), (17, 19), (29, 31), (41, 43), (59, 61), (71, 73), (101, 103), (107, 109), (137, 139), … A077800.

Every twin prime pair except (3, 5) is of the form (6n − 1, 6n + 1) for some natural number n; that is, the number between the two primes is a multiple of 6.

Brun's Theorem

In 1915, Viggo Brun showed that the sum of reciprocals of the twin primes was convergent. This famous result, called Brun's theorem, was the first use of the Brun sieve and helped initiate the development of modern sieve theory. The modern version of Brun's argument can be used to show that the number of twin primes less than N does not exceed

$$\frac{CN}{(\log N)^2}$$

for some absolute constant C > 0. In fact, it is bounded above by

$\frac{C'N}{(\log N)^2}\left(1+O\left(\frac{\log\log N}{\log N}\right)\right)$ where C' = $8C_2$, where C_2 is the **twin prime constant**, given below.

Other Theorems Weaker than the Twin Prime Conjecture

In 1940, Paul Erdős showed that there is a constant c < 1 and infinitely many primes p such that (p′ − p) < (c ln p) where p′ denotes the next prime after p. This result was successively improved; in 1986 Helmut Maier showed that a constant c < 0.25 can be used. In 2004 Daniel Goldston and Cem Yıldırım showed that the constant could be improved further to c = 0.085786... In 2005, Goldston, János Pintz and Yıldırım established that c can be chosen to be arbitrarily small

$$\liminf_{n\to\infty}\frac{p_{n+1}-p_n}{\log p_n}=0.$$

By assuming the Elliott–Halberstam conjecture or a slightly weaker version, they were able to show that there are infinitely many n such that at least two of n, n + 2, n + 6, n + 8, n + 12, n + 18, or n + 20 are prime. Under a stronger hypothesis they showed that for infinitely many n, at least two of n, n + 2, n + 4, and n + 6 are prime. The result of Yitang Zhang,

$$\liminf_{n\to\infty}(p_{n+1}-p_n)<N \text{ with } N=7\times 10^7,$$

is a major improvement on the Goldston–Graham–Pintz–Yıldırım result. The Polymath Project optimization of Zhang's bound and Maynard claims to have reduced the bound to N = 246 are further improvements.

Sequence

The twin prime conjecture can be proved when the sequenceA165959 has been shown to have an infinite number of terms which equal 3. This sequence is base on the difference of both sides of the Ramanujan prime corollary. This corollary states:

$$2p_{i-n}>p_i \text{ for } i>k \text{ where } k=\pi(p_k)=\pi(R_n),$$

where $\pi(x)$ is the prime-counting function, equal to the number of primes less than or equal to x. The corollary is very useful in showing the number of primes in the range $[p_k, 2p_{i-n}]$ and if the size of this range is 3, only $p_k + 2$ the larger twin prime to p_k can exist.

Conjectures

First Hardy–Littlewood Conjecture

The **Hardy–Littlewood conjecture** (named after G. H. Hardy and John Littlewood) is a generalization of the twin prime conjecture. It is concerned with the distribution of prime constellations, including twin primes, in analogy to the prime number theorem. Let $\pi_2(x)$ denote the number of primes $p \leq x$ such that $p + 2$ is also prime. Define the **twin** prime constant C_2 as

$$C_2 = \prod_{\substack{p \text{ prime} \\ p \geq 3}} \left(1 - \frac{1}{(p-1)^2}\right) \approx 0.66016181584686957392781211001 4\ldots$$

A005597 (here the product extends over all prime numbers $p \geq 3$). Then the conjecture is that

$$\pi_2(x) \sim 2C_2 \frac{x}{(\ln x)^2} \sim 2C_2 \int_2^x \frac{dt}{(\ln t)^2}$$

in the sense that the quotient of the two expressions tends to 1 as x approaches infinity. (The second ~ is not part of the conjecture and is proved by integration by parts.)

The conjecture can be justified (but not proven) by assuming that 1 / ln t describes the density function of the prime distribution, an assumption suggested by the prime number theorem and would imply the twin prime conjecture, but remains unresolved.

This first Hardy–Littlewood conjecture on prime k-tuples implies that the second Hardy–Littlewood conjecture is false.

Polignac's Conjecture

Polignac's conjecture from 1849 states that for every positive even natural number k, there are infinitely many consecutive prime pairs p and p such that $p' - p = k$ (i.e. there are infinitely many prime gaps of size k). The case $k = 2$ is the twin prime conjecture. The conjecture has not yet been proven or disproven for any specific value of k, but Zhang's result proves that it is true for at least one (currently unknown) value of k. Indeed, if such a k did not exist, then for any positive even natural number N there are at most finitely many n such that $p_{n+1} - p_n = m$ for all $m < N$ and so for n large enough we have $p_{n+1} - p_n > N$, which would contradict Zhang's result.

Large Twin Primes

Beginning in 2007, two distributed computing projects, Twin Prime Search and PrimeGrid, have produced several record-largest twin primes. As of September 2016, the current largest twin prime pair known is $2996863034895 \cdot 2^{1290000} \pm 1$, with 388,342 decimal digits. It was discovered in September 2016.

There are 808,675,888,577,436 twin prime pairs below 10^{18}.

An empirical analysis of all prime pairs up to $4.35 \cdot 10^{15}$ shows that if the number of such pairs less than x is $f(x) \cdot x/(\log x)^2$ then f(x) is about 1.7 for small x and decreases towards about 1.3 as x tends to infinity. The limiting value of f(x) is conjectured to equal twice the twin prime constant (OEIS A114907) (not to be confused with Brun's constant), according to the Hardy–Littlewood conjecture.

Other Elementary Properties

Every third odd number is divisible by 3, which requires that no three successive odd numbers can be prime unless one of them is 3. Five is therefore the only prime that is part of two twin prime pairs. The lower member of a pair is by definition a Chen prime.

It has been proven that the pair (m, m + 2) is a twin prime if and only if

$$4((m-1)!+1) \equiv -m \pmod{m(m+2)}.$$

If m − 4 or m + 6 is also prime then the three primes are called a prime triplet.

For a twin prime pair of the form (6n − 1, 6n + 1) for some natural number n > 1, n must have units digit 0, 2, 3, 5, 7, or 8 (A002822).

The sum of twin primes greater than 3 is divisible by 12.

Isolated Prime

An isolated prime (also known as single prime or non-twin prime) is a prime number p such that neither p − 2 nor p + 2 is prime. In other words, p is not part of a twin prime pair. For example, 23 is an isolated prime, since 21 and 25 are both composite.

The first few isolated primes are

2, 23, 37, 47, 53, 67, 79, 83, 89, 97, ... A007510

It follows from Brun's theorem that almost all primes are members of this sequence.

COLLATZ CONJECTURE

The **Collatz conjecture** is a conjecture in mathematics named after Lothar Collatz. The conjecture is also known as the **3*n* + 1 conjecture**, the **Ulam conjecture** (after Stanisław Ulam), **Kakutani's problem** (after Shizuo Kakutani), the **Thwaites conjecture** (after Sir Bryan Thwaites), **Hasse's algorithm** (after Helmut Hasse), or the **Syracuse problem**; the sequence of numbers involved is referred to as the **hailstone sequence** or **hailstone**

numbers (because the values are usually subject to multiple descents and ascents like hailstones in a cloud), or as **wondrous numbers**.

The conjecture can be summarized as follows. Take any positive integer n. If n is even, divide it by 2 to get n / 2. If n is odd, multiply it by 3 and add 1 to obtain 3n + 1. Repeat the process (which has been called "Half Or Triple Plus One, or **HOTPO**) indefinitely. The conjecture is that no matter what number you start with, you will always eventually reach 1.

Paul Erdős said about the Collatz conjecture: "Mathematics may not be ready for such problems." He also offered $500 for its solution. Jeffrey Lagarias in 2010 claimed that based only on known information about this problem, "this is an extraordinarily difficult problem, completely out of reach of present day mathematics."

Statement of the Problem

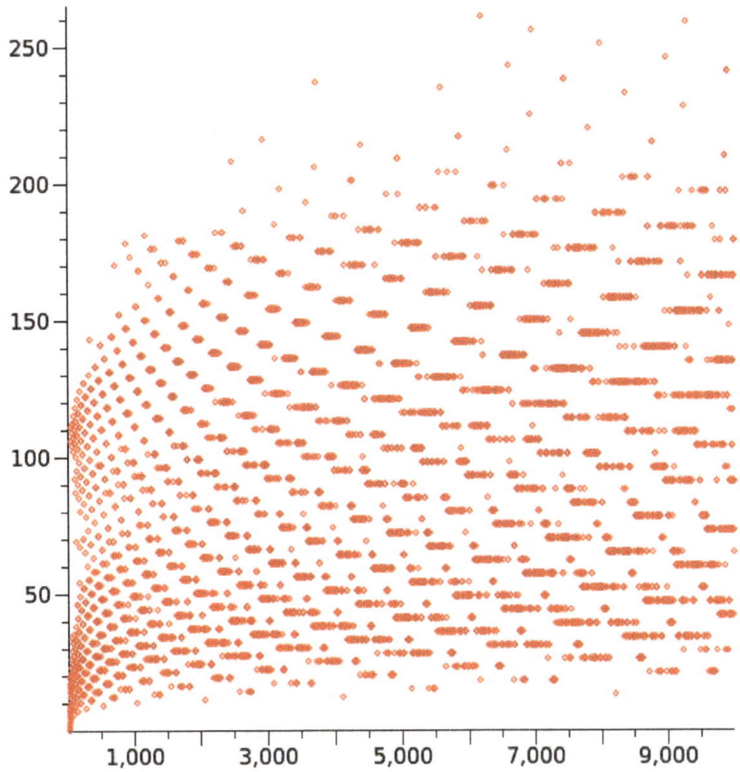

Numbers from 1 to 9999 and their corresponding total stopping time

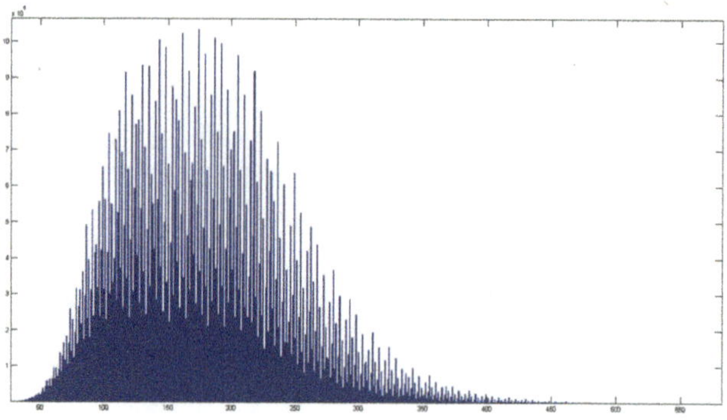

Histogram of total stopping times for the numbers 1 to 100 million. Total stopping time is on the x axis, frequency on the y axis. Note that for all positive integers the histogram would be a completely different, exponentially-growing sequence (see #In reverse)

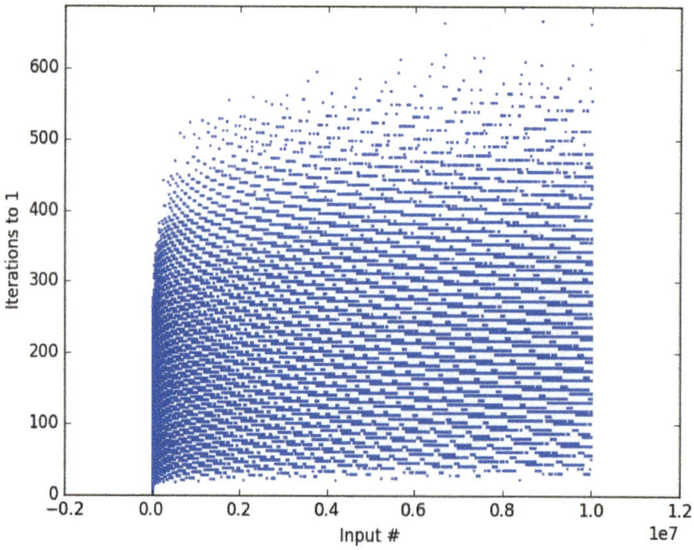

Iteration time for inputs of 2 to 10,000,000

Consider the following operation on an arbitrary positive integer:

- If the number is even, divide it by two.
- If the number is odd, triple it and add one.

In modular arithmetic notation, define the function f as follows:

$$f(n) = \begin{cases} n/2 & \text{if } n \equiv 0 \pmod 2 \\ 3n+1 & \text{if } n \equiv 1 \pmod 2). \end{cases}$$

Now, form a sequence by performing this operation repeatedly, beginning with any positive integer, and taking the result at each step as the input at the next.

In notation:

$$a_i = \begin{cases} n & \text{for } i = 0 \\ f(a_{i-1}) & \text{for } i > 0 \end{cases}$$

(that is: a_i is the value of f applied to n recursively i times; $a_i = f^i(n)$).

The Collatz conjecture is: This process will eventually reach the number 1, regardless of which positive integer is chosen initially.

That smallest i such that $a_i = 1$ is called the **total stopping time** of n. The conjecture asserts that every n has a well-defined total stopping time. If, for some n, such an i doesn't exist, we say that n has infinite total stopping time and the conjecture is false.

If the conjecture is false, it can only be because there is some starting number which gives rise to a sequence that does not contain 1. Such a sequence might enter a repeating cycle that excludes 1, or increase without bound. No such sequence has been found.

Examples

For instance, starting with n = 12, one gets the sequence 12, 6, 3, 10, 5, 16, 8, 4, 2, 1.

n = 19, for example, takes longer to reach 1: 19, 58, 29, 88, 44, 22, 11, 34, 17, 52, 26, 13, 40, 20, 10, 5, 16, 8, 4, 2, 1.

The sequence for n = 27, listed and graphed below, takes 111 steps (41 steps through **odd numbers**), climbing to 9232 before descending to 1.

27, 82, **41**, 124, 62, **31**, 94, **47**, 142, **71**, 214, **107**, 322, **161**, 484, 242, **121**, 364, 182, **91**, 274, **137**, 412, 206, **103**, 310, **155**, 466, **233**, 700, 350, **175**, 526, **263**, 790, **395**, 1186, **593**, 1780, 890, **445**, 1336, 668, 334, **167**, 502, **251**, 754, **377**, 1132, 566, **283**, 850, **425**, 1276, 638, **319**, 958, **479**, 1438, **719**, 2158, **1079**, 3238, **1619**, 4858, **2429**, 7288, 3644, 1822, **911**, 2734, **1367**, 4102, **2051**, 6154, **3077**, 9232, 4616, 2308, 1154, **577**, 1732, 866, **433**, 1300, 650, **325**, 976, 488, 244, 122, **61**, 184, 92, 46, **23**, 70, **35**, 106, **53**, 160, 80, 40, 20, 10, **5**, 16, 8, 4, 2, **1** (sequence A008884 in the OEIS)

1, 2, 3, 6, 7, 9, 18, 25, 27, 54, 73, 97, 129, 171, 231, 313, 327, 649, 703, 871, 1161, 2223, 2463, 2919, 3711, 6171, ... (sequence A006877 in the OEIS).

The starting values whose maximum trajectory point is greater than that of any smaller starting value are as follows:

1, 2, 3, 7, 15, 27, 255, 447, 639, 703, 1819, 4255, 4591, 9663, 20895, 26623, 31911, 60975, 77671, 113383, 138367, 159487, 270271, 665215, 704511, ... (sequence A006884 in the OEIS)

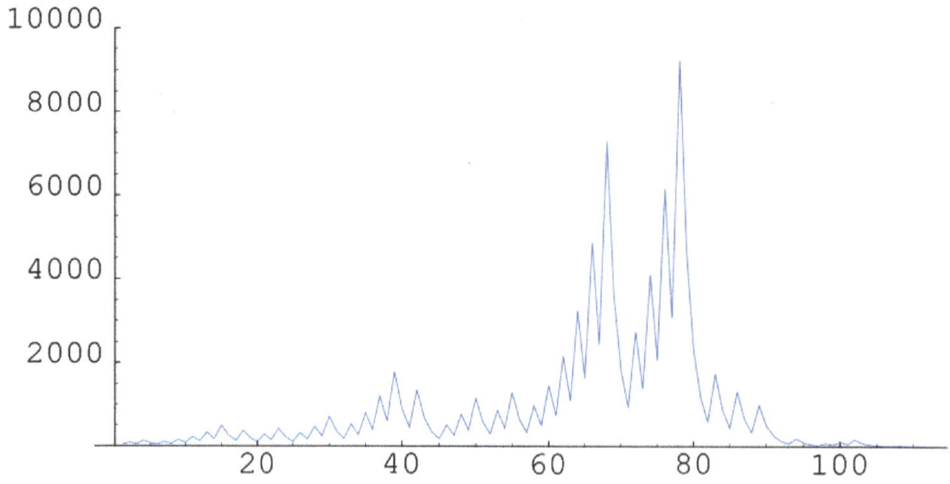

Numbers with a total stopping time longer than any smaller starting value form a sequence beginning with:

Number of steps for n to reach 1 are

0, 1, 7, 2, 5, 8, 16, 3, 19, 6, 14, 9, 9, 17, 17, 4, 12, 20, 20, 7, 7, 15, 15, 10, 23, 10, 111, 18, 18, 18, 106, 5, 26, 13, 13, 21, 21, 21, 34, 8, 109, 8, 29, 16, 16, 16, 104, 11, 24, 24, ... (sequence A006577 in the OEIS)

The longest progression for any initial starting number

less than 10 is 9, which has 19 steps,

less than 100 is 97, which has 118 steps,

less than 1,000 is 871, which has 178 steps,

less than 10,000 is 6,171, which has 261 steps,

less than 100,000 is 77,031, which has 350 steps,

less than 1 million is 837,799, which has 524 steps,

less than 10 million is 8,400,511, which has 685 steps,

less than 100 million is 63,728,127, which has 949 steps,

less than 1 billion is 670,617,279, which has 986 steps,

less than 10 billion is 9,780,657,631, which has 1132 steps, and

less than 100 billion is 75,128,138,247, which has 1228 steps.

The powers of two converge to one quickly because 2^n is halved n times to reach one, and is never increased, but for Mersenne number M_n, they need to increase n times and usually need more steps to reach 1.

Visualizations

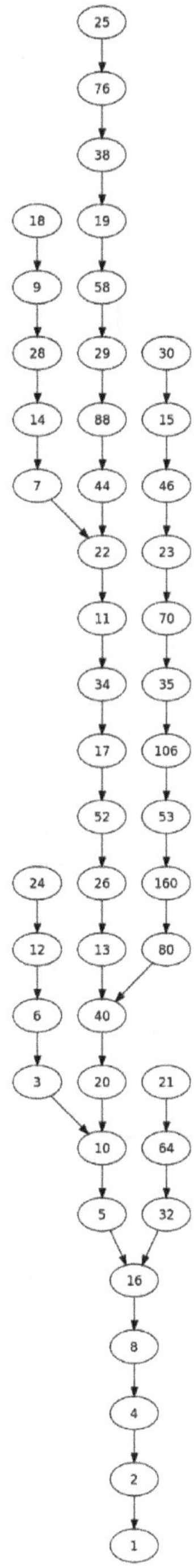

Directed graph showing the orbits of small numbers under the Collatz map. The Collatz conjecture is equivalent to the statement that all paths eventually lead to 1.

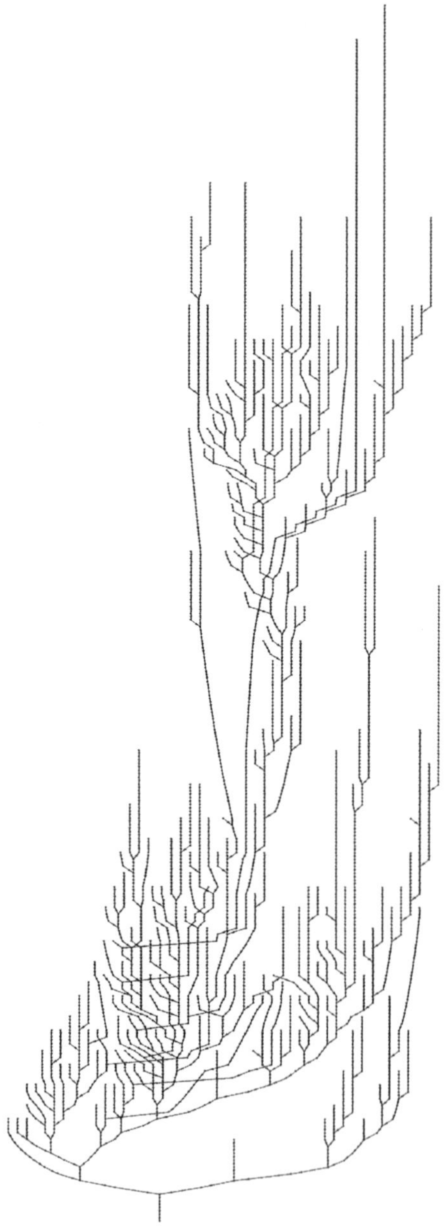

Directed graph showing the orbits of the first 1000 numbers.

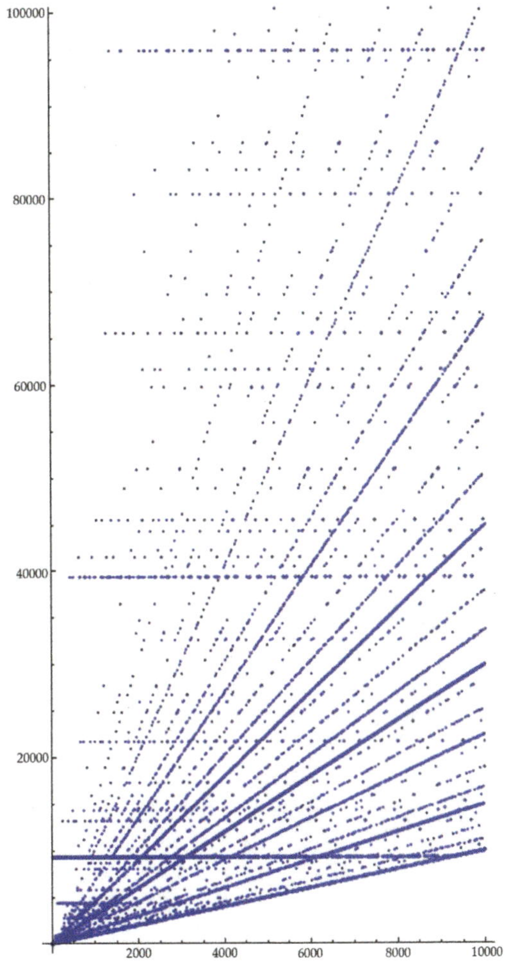

The x axis represents starting number, the y axis represents the highest number reached during the chain to 1.

Cycles

Any counterexample to the Collatz conjecture would have to consist either of an infinite divergent trajectory or a cycle different from the trivial (4; 2; 1) cycle. Thus, if one could prove that neither of these types of counterexample could exist, then all positive integers would have a trajectory that reaches the trivial cycle. Such a strong result is not known, but certain types of cycles have been ruled out.

The type of a cycle may be defined with reference to the "shortcut" definition of the Collatz map, $f(n) = (3n+1)/2$ for odd *n* and $f(n) = n/2$ for even n. A cycle is a sequence $(a_0; a_1; \dots a_q)$ where $f(a_0) = a_1$, $f(a_1) = a_2$, and so on, up to $f(a_q) = a_0$ in a closed loop. For this shortcut definition, the only known cycle is (1; 2). Although 4 is part of the single known cycle for the original Collatz map, it is not part of the cycle for the shortcut map.

A k-cycle is a cycle that can be partitioned into 2k contiguous subsequences: k increasing sequences of odd numbers alternating with k decreasing sequences of even numbers. For instance, if the cycle consists of a single increasing sequence of odd numbers followed by a decreasing sequence of even numbers, it is called a 1-cycle.

Steiner (1977) proved that there is no 1-cycle other than the trivial (1;2). Simons (2004) used Steiner's method to prove that there is no 2-cycle. Simons & de Weger (2003) extended this proof up to 68-cycles: there is no k-cycle up to k = 68. Beyond 68, this method gives upper bounds for the elements in such a cycle: for example, if there is a 75-cycle, then at least one element of the cycle is less than 2385×2^{50}.Therefore, as exhaustive computer searches continue, larger cycles may be ruled out. To state the argument more intuitively: we need not look for cycles that have at most 68 trajectories, where each trajectory consists of consecutive ups followed by consecutive downs. See below for an idea of how one might find an upper bound for the elements of a cycle.

Supporting Arguments

Although the conjecture has not been proven, most mathematicians who have looked into the problem think the conjecture is true because experimental evidence and heuristic arguments support it.

Experimental Evidence

The conjecture has been checked by computer for all starting values up to 2^{60}. All initial values tested so far eventually end in the repeating cycle (4; 2; 1), which has only three terms. From this lower bound on the starting value, a lower bound can also be obtained for the number of terms a repeating cycle other than (4; 2; 1) must have. When this relationship was established in 1981, the formula gave a lower bound of 35,400 terms.

This computer evidence is not a proof that the conjecture is true. As shown in the cases of the Pólya conjecture, the Mertens conjecture and the Skewes' number, sometimes a conjecture's only counterexamples are found when using very large numbers.

A Probabilistic Heuristic

If one considers only the odd numbers in the sequence generated by the Collatz process, then each odd number is on average 3/4 of the previous one. (More precisely, the geometric mean of the ratios of outcomes is 3/4.) This yields a heuristic argument that every Hailstone sequence should decrease in the long run, although this is not evidence against other cycles, only against divergence. The argument is not a proof because it assumes that Hailstone sequences are assembled from uncorrelated probabilistic events. (It does rigorously establish that the 2-adic extension of the Collatz process has two division steps for every multiplication step for almost all 2-adic starting values.)

And even if the probabilistic reasoning were rigorous, this would still imply only that the conjecture is almost surely true for any given integer, which does not necessarily imply that it is true for all integers.

Rigorous Bounds

Although it is not known rigorously whether all positive numbers eventually reach one according to the Collatz iteration, it is known that many numbers do so. In particular, Krasikov and Lagarias showed that the number of integers in the interval [1,x] that eventually reach one is at least proportional to $x^{0.84}$.

Other Formulations of the Conjecture

In Reverse

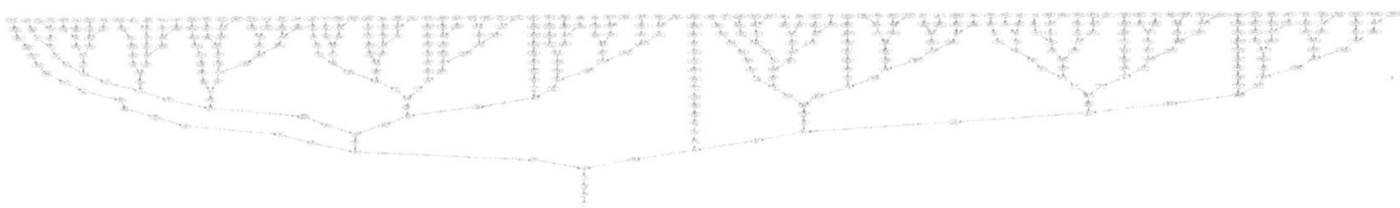

The first 21 levels of the Collatz graph generated in bottom-up fashion. The graph includes all numbers with an orbit length of 21 or less.

There is another approach to prove the conjecture, which considers the bottom-up method of growing the so-called Collatz graph. The Collatz graph is a graph defined by the inverse relation

$$R(n) = \begin{cases} \{2n\} & \text{if } n \equiv 0,1,2,3,5 \\ \{2n, (n-1)/3\} & \text{if } n \equiv 4 \end{cases} \pmod{6}.$$

So, instead of proving that all positive integers eventually lead to 1, we can try to prove that 1 leads to all positive integers. For any integer n, n ≡ 1 (mod 2) iff 3n + 1 ≡ 4 (mod 6). Equivalently, (n − 1)/3 ≡ 1 (mod 2) iff n ≡ 4 (mod 6). Conjecturally, this inverse relation forms a tree except for the 1–2–4 loop (the inverse of the 4–2–1 loop of the unaltered function f defined in the Statement of the problem section of this article).

When the relation 3n + 1 of the function f is replaced by the common substitute “shortcut” relation (3n + 1)/2, the Collatz graph is defined by the inverse relation,

$$R(n) = \begin{cases} \{2n\} & \text{if } n \equiv 0,1 \\ \{2n, (2n-1)/3\} & \text{if } n \equiv 2 \end{cases} \pmod{3}.$$

For any integer n, n ≡ 1 (mod 2) iff (3n + 1)/2 ≡ 2 (mod 3). Equivalently, (2n − 1)/3 ≡ 1 (mod 2) iff n ≡ 2 (mod 3). Conjecturally, this inverse relation forms a tree except for a 1–2 loop (the inverse of the 1–2 loop of the function f(n) revised as indicated above).

Alternately, replace the 3n + 1 with n' / H(n') where n' = 3n + 1 and H(n') is the highest power of 2 that divides n' (with no remainder). The resulting function f maps from odd numbers to odd numbers. Now suppose that for some odd number n, applying this operation k times yields the number 1 (that is, $f^k(n) = 1$). Then in binary, the number n can be written as the concatenation of strings $w_k\ w_{k-1} \ldots w_1$ where each w_h is a finite and contiguous extract from the representation of $1 / 3^h$. The representation of n therefore holds the repetends of $1 / 3^h$, where each repetend is optionally rotated and then re0plicated up to a finite number of bits. It is only in binary that this occurs. Conjecturally, every binary string s that ends with a '1' can be reached by a representation of this form (where we may add or delete leading '0's to s).

As an Abstract Machine that Computes in Base Two

Repeated applications of the Collatz function can be represented as an abstract machine that handles strings of bits. The machine will perform the following three steps on any odd number until only one "1" remains:

- Append 1 to the (right) end of the number in binary (giving 2n + 1);
- Add this to the original number by binary addition (giving 2n + 1 + n = 3n + 1);
- Remove all trailing 0s (i.e. repeatedly divide by two until the result is odd).

This prescription is plainly equivalent to computing a Hailstone sequence in base two.

Example

The starting number 7 is written in base two as 111. The resulting Hailstone sequence is:

```
       111
      1111
     10110
    10111
   100010
  100011
  110100
 11011
101000
1011
10000
```

As a Parity Sequence

For this section, consider the Collatz function in the slightly modified form

$$f(n) = \begin{cases} n/2 & \text{if } n \equiv 0 \\ (3n+1)/2 & \text{if } n \equiv 1. \end{cases} \pmod{2}$$

This can be done because when n is odd, 3n + 1 is always even.

If P(...) is the parity of a number, that is P(2n) = 0 and P(2n + 1) = 1, then we can define the Hailstone parity sequence (or parity vector) for a number n as $p_i = P(a_i)$, where $a_0 = n$, and $a_{i+1} = f(a_i)$.

What operation is performed (3n + 1)/2 or n/2 depends on the parity. The parity sequence is the same as the sequence of operations.

Using this form for f(n), it can be shown that the parity sequences for two numbers m and n will agree in the first k terms if and only if m and n are equivalent modulo 2^k. This implies that every number is uniquely identified by its parity sequence, and moreover that if there are multiple Hailstone cycles, then their corresponding parity cycles must be different.

Applying the f function k times to the number $a \cdot 2^k + b$ will give the result $a \cdot 3^c + d$, where d is the result of applying the f function k times to b, and c is how many odd numbers were encountered during that sequence.

As a Tag System

For the Collatz function in the form

$$f(n) = \begin{cases} n/2 & \text{if } n \equiv 0 \\ (3n+1)/2 & \text{if } n \equiv 1. \end{cases} \pmod{2}$$

Hailstone sequences can be computed by the extremely simple 2-tag system with production rules a → bc, b → a, c → aaa. In this system, the positive integer n is represented by a string of n a, and iteration of the tag operation halts on any word of length less than 2. (Adapted from De Mol.)

The Collatz conjecture equivalently states that this tag system, with an arbitrary finite string of a's as the initial word, eventually halts (see Tag system#Example: Computation of Collatz sequences for a worked example).

Extensions to Larger Domains

Iterating on all Integers

An extension to the Collatz conjecture is to include all integers, not just positive integers. Leaving aside the cycle 0 → 0, there are a total of 4 known cycles, which all nonzero integers seem to eventually fall into under iteration of f. These cycles are listed here, starting with the well-known cycle for positive n:

Odd values are listed in large bold. Each cycle is listed with its member of least absolute value (which is always odd) first.

Cycle	Odd-value cycle length	Full cycle length
1 → 4 → 2 → **1** ...	1	3
−1 → −2 → **−1** ...	1	2
−5 → −14 → **−7** → −20 → −10 → **−5** ...	2	5
−17 → −50 → **−25** → −74 → **−37** → −110 → **−55** → −164 → −82 → **−41** → −122 → **−61** → −182 → **−91** → −272 → −136 → −68 → −34 → **−17** ...	7	18

The generalized Collatz conjecture is the assertion that every integer, under iteration by f, eventually falls into one of the four cycles above or the cycle 0 → 0. The 0 → 0 cycle is often regarded as "trivial" by the argument, as it is only included for the sake of completeness.

Iterating with Odd Denominators or 2-Adic Integers

The standard Collatz map can be extended to (positive or negative) rational numbers which have odd denominators when written in lowest terms. The number is taken to be odd or even according to whether its numerator is odd or even. A closely related fact is that the Collatz map extends to the ring of 2-adic integers, which contains the ring of rationals with odd denominators as a subring.

The parity sequences as defined above are no longer unique for fractions. However, it can be shown that any possible parity cycle is the parity sequence for exactly one fraction: if a cycle has length n and includes odd numbers exactly m times at indices k_0, ..., k_{m-1}, then the unique fraction which generates that parity cycle is

$$\frac{3^{m-1}2^{k_0}+\cdots+3^0 2^{k_{m-1}}}{2^n-3^m}.$$

For example, the parity cycle (1 0 1 1 0 0 1) has length 7 and has 4 odd numbers at indices 0, 2, 3, and 6. The unique fraction which generates that parity cycle is

$$\frac{3^3 2^0+3^2 2^2+3^1 2^3+3^0 2^6}{2^7-3^4}=\frac{151}{47},$$

the complete cycle being: 151/47 → 250/47 → 125/47 → 211/47 → 340/47 → 170/47 → 85/47 → 151/47

Although the cyclic permutations of the original parity sequence are unique fractions, the cycle is not unique, each permutation's fraction being the next number in the loop cycle:

$$(0\,1\,1\,0\,0\,1\,1) \rightarrow \frac{3^3 2^1 + 3^2 2^2 + 3^1 2^5 + 3^0 2^6}{2^7 - 3^4} = \frac{250}{47}$$

$$(1\,1\,0\,0\,1\,1\,0) \rightarrow \frac{3^3 2^0 + 3^2 2^1 + 3^1 2^4 + 3^0 2^5}{2^7 - 3^4} = \frac{125}{47}$$

$$(1\,0\,0\,1\,1\,0\,1) \rightarrow \frac{3^3 2^0 + 3^2 2^3 + 3^1 2^4 + 3^0 2^6}{2^7 - 3^4} = \frac{211}{47}$$

$$(0\,0\,1\,1\,0\,1\,1) \rightarrow \frac{3^3 2^2 + 3^2 2^3 + 3^1 2^5 + 3^0 2^6}{2^7 - 3^4} = \frac{340}{47}$$

$$(0\,1\,1\,0\,1\,1\,0) \rightarrow \frac{3^3 2^1 + 3^2 2^2 + 3^1 2^4 + 3^0 2^5}{2^7 - 3^4} = \frac{170}{47}$$

$$(1\,1\,0\,1\,1\,0\,0) \rightarrow \frac{3^3 2^0 + 3^2 2^1 + 3^1 2^3 + 3^0 2^4}{2^7 - 3^4} = \frac{85}{47}$$

Also, for uniqueness, the parity sequence should be “prime”, i.e., not partitionable into identical sub-sequences. For example, parity sequence (1 1 0 0 1 1 0 0) can be partitioned into two identical sub-sequences (1 1 0 0)(1 1 0 0). Calculating the 8-element sequence fraction gives

$$(1\,1\,0\,0\,1\,1\,0\,0) \rightarrow \frac{3^3 2^0 + 3^2 2^1 + 3^1 2^4 + 3^0 2^5}{2^8 - 3^4} = \frac{125}{175}$$

But when reduced to lowest terms {5/7}, it is the same as that of the 4-element sub-sequence

$$(1\,1\,0\,0) \rightarrow \frac{3^1 2^0 + 3^0 2^1}{2^4 - 3^2} = \frac{5}{7}.$$

And this is because the 8-element parity sequence actually represents two circuits of the loop cycle defined by the 4-element parity sequence.

In this context, the Collatz conjecture is equivalent to saying that (0 1) is the only cycle which is generated by positive whole numbers (i.e. 1 and 2).

Cycle Bounds

Eq. (1) also gives a rough idea about how one can prove that cycles of certain lengths do not exist. For a hypothetical cycle of length n, the numerator is bounded above by $3^n - 2^n$ (this corresponds to a cycle of all odd numbers). A lower bound for the denominator can be obtained by letting n/m be an optimal rational approximation to log(3)/log(2). Together these give an upper bound for the unique fraction that generates a cycle of length n. If this upper bound is smaller than the largest number for which the conjecture has been verified to hold, then a cycle of length n is impossible.

Iterating on Real or Complex Numbers

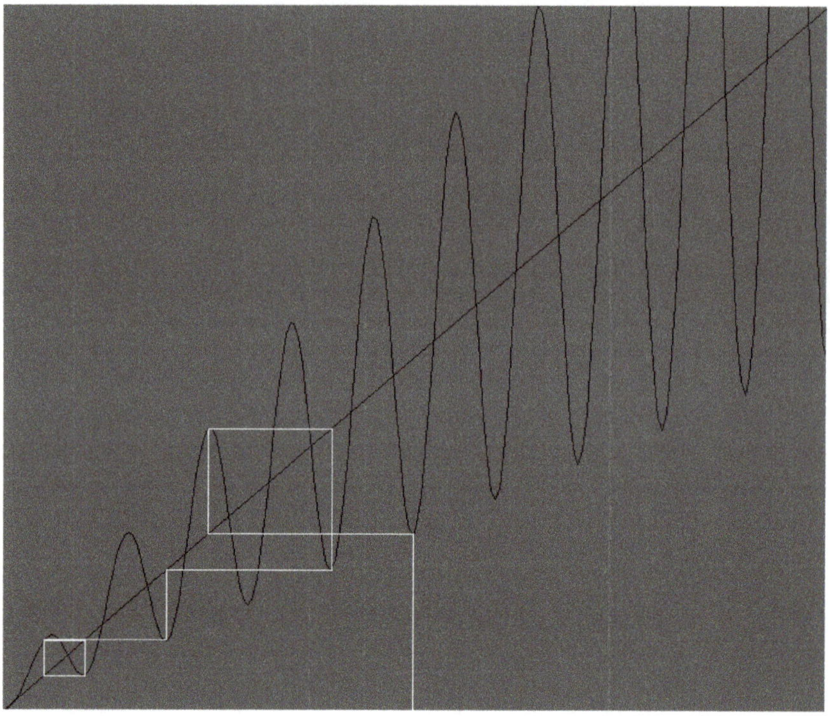

Cobweb plot of the orbit 10-5-8-4-2-1-2-1-2-1-etc. in the real extension of the Collatz map (optimized by replacing "3n + 1" with "(3n + 1)/2")

The Collatz map can be viewed as the restriction to the integers of the smooth real and complex map

$$f(z) = \frac{1}{2} z \cos^2\left(\frac{\pi}{2} z\right) + (3z+1)\sin^2\left(\frac{\pi}{2} z\right),$$

which simplifies to $\frac{1}{4}(2 + 7z - (2 + 5z)\cos(\pi z))$.

If the standard Collatz map defined above is optimized by replacing the relation 3n + 1 with the common substitute "shortcut" relation (3n + 1)/2, it can be viewed as the restriction to the integers of the smooth real and complex map

$$f(z) = \frac{1}{2} z \cos^2\left(\frac{\pi}{2} z\right) + \frac{(3z+1)}{2}\sin^2\left(\frac{\pi}{2} z\right),$$

which simplifies to $\frac{1}{4}(1 + 4z - (1 + 2z)\cos(\pi z))$.

Collatz Fractal

Collatz map fractal in a neighbourhood of the real line

Iterating the above optimized map in the complex plane produces the Collatz fractal.

The point of view of iteration on the real line was investigated by Chamberland (1996), and on the complex plane by Letherman, Schleicher, and Wood (1999).

Optimizations

Time-Space Tradeoff

The section As a parity sequence above gives a way to speed up simulation of the sequence. To jump ahead k steps on each iteration (using the f function from that section), break up the current number into two parts, b (the k least significant bits, interpreted as an integer), and a (the rest of the bits as an integer). The result of jumping ahead k steps can be found as:

$f^{k}(a\ 2^{k} + b) = a\ 3^{c(b)} + d(b)$.

The c (or better 3^c) and d arrays are precalculated for all possible k-bit numbers b, where d(b) is the result of applying the f function k times to b, and c(b) is the number of odd numbers encountered on the way. For example, if k=5, you can jump ahead 5 steps on each iteration by separating out the 5 least significant bits of a number and using:

c(0..31) = {0,3,2,2,2,2,2,4,1,4,1,3,2,2,3,4,1,2,3,3,1,1,3,3,2,3,2,4,3,3,4,5}

d(0..31) = {0,2,1,1,2,2,2,20,1,26,1,10,4,4,13,40,2,5,17,17,2,2,20,20,8,22,8,71,26,26,80,242}.

This requires 2^k precomputation and storage to speed up the resulting calculation by a factor of k, a space-time tradeoff.

Modular Restrictions

For the special purpose of searching for a counterexample to the Collatz conjecture, this precomputation leads to an even more important acceleration, used by Tomás Oliveira e Silva in his computational confirmations of the Collatz conjecture up to large values of n. If, for some given b and k, the inequality

$f^{k}(a\,2^k + b) = a\,3^{c(b)} + d(b) < a\,2^k + b$

holds for all a, then the first counterexample, if it exists, cannot be b modulo 2^k. For instance, the first counterexample must be odd because f(2n) = n, smaller than 2n; and it must be 3 mod 4 because $f^2(4n + 1) = 3n + 1$, smaller than 4n + 1. For each starting value a which is not a counterexample to the Collatz conjecture, there is a k for which such an inequality holds, so checking the Collatz conjecture for one starting value is as good as checking an entire congruence class. As k increases, the search only needs to check those residues b that are not eliminated by lower values of k. Only an exponentially small fraction of the residues survive. For example, the only surviving residues mod 32 are 7, 15, 27, and 31.

Syracuse Function

If k is an odd integer, then $3k+1$ is even, so $3k+1=2^a k'$ with k' odd and $a \geq 1$. The **Syracuse function** is the function f from the set I of odd integers into itself, for which $f(k)=k'$ (sequence A075677 in the OEIS).

Some properties of the Syracuse function are:

$\forall k \in I, f(4k+1)=f(k)$. (Because $3(4k+1)+1=12k+4=4(3k+1)$)

- In more generality: For all $p \geq 1$ and odd h, $f^{p-1}(2^p h-1)=2\cdot 3^{p-1}h-1$. (Here f^{p-1} is function iteration notation.)
- For all odd $h, f(2h-1) \leq \frac{3h-1}{2}$

The Collatz conjecture is equivalent to the statement that, for all k in I, there exists an integer $n \geq 1$ such that $f^n(k)=1$.

Undecidable Generalizations

In 1972, J. H. Conway proved that a natural generalization of the Collatz problem is algorithmically undecidable.

Specifically, he considered functions of the form

$$g(n)=a_i n+b_i, n \equiv i \pmod P$$

Where $a_0, b_0, \cdots, a_{P-1}, b_{P-1}$ are rational numbers which are so chosen that $g(n)$ is always integral.

The standard Collatz function is given by $P = 2,\ a_0 = \frac{1}{2},\ b_0 = 0,\ a_1 = 3,\ b_1 = 1$. Conway proved that the problem:

Given g and n, does the sequence of iterates $g^k(n)$ reach 1? is undecidable, by representing the halting problem in this way. Closer to the Collatz problem is the following universally quantified problem:

Given g does the sequence of iterates $g^k(n)$ reach 1, *for all* $n > 0$?

Modifying the condition in this way can make a problem either harder or easier to solve (intuitively, it is harder to justify a positive answer but might be easier to justify a negative one). Kurtz and Simon proved that the above problem is, in fact, undecidable and even higher in the arithmetical hierarchy, specifically Π_2^0-complete. This hardness result holds even if one restricts the class of functions g by fixing the modulus P to 6480.

AdS/CFT Correspondence

In theoretical physics, the **anti-de Sitter/conformal field theory correspondence**, sometimes called **Maldacena duality** or **gauge/gravity duality**, is a conjectured relationship between two kinds of physical theories. On one side are anti-de Sitter spaces (AdS) which are used in theories of quantum gravity, formulated in terms of string theory or M-theory. On the other side of the correspondence are conformal field theories (CFT) which are quantum field theories, including theories similar to the Yang–Mills theories that describe elementary particles.

The duality represents a major advance in our understanding of string theory and quantum gravity.This is because it provides a non-perturbative formulation of string theory with certain boundary conditions and because it is the most successful realization of the holographic principle, an idea in quantum gravity originally proposed by Gerard 't Hooft and promoted by Leonard Susskind.

It also provides a powerful toolkit for studying strongly coupled quantum field theories. Much of the usefulness of the duality results from the fact that it is a strong-weak duality: when the fields of the quantum field theory are strongly interacting, the ones in the gravitational theory are weakly interacting and thus more mathematically tractable. This fact has been used to study many aspects of nuclear and condensed matter physics by translating problems in those subjects into more mathematically tractable problems in string theory.

The AdS/CFT correspondence was first proposed by Juan Maldacena in late 1997. Important aspects of the correspondence were elaborated in articles by Steven Gubser, Igor Klebanov, and Alexander Markovich Polyakov, and by Edward Witten. By 2015, Maldacena's article had over 10,000 citations, becoming the most highly cited article in the field of high energy physics.

Background

Quantum Gravity and Strings

Our current understanding of gravity is based on Albert Einstein's general theory of relativity. Formulated in 1915, general relativity explains gravity in terms of the geometry of space and time, or spacetime. It is formulated in the language of classical physics developed by physicists such as Isaac Newton and James Clerk Maxwell. The other nongravitational forces are explained in the framework of quantum mechanics. Developed in the first half of the twentieth century by a number of different physicists, quantum mechanics provides a radically different way of describing physical phenomena based on probability.

Quantum gravity is the branch of physics that seeks to describe gravity using the principles of quantum mechanics. Currently, the most popular approach to quantum gravity is string theory, which models elementary particles not as zero-dimensional points but as one-dimensional objects called strings. In the AdS/CFT correspondence, one typically considers theories of quantum gravity derived from string theory or its modern extension, M-theory.

In everyday life, there are three familiar dimensions of space (up/down, left/right, and forward/backward), and there is one dimension of time. Thus, in the language of modern physics, one says that spacetime is four-dimensional. One peculiar feature of string theory and M-theory is that these theories require extra dimensions of spacetime for their mathematical consistency: in string theory spacetime is ten-dimensional, while in M-theory it is eleven-dimensional. The quantum gravity theories appearing in the AdS/CFT correspondence are typically obtained from string and M-theory by a process known as compactification. This produces a theory in which spacetime has effectively a lower number of dimensions and the extra dimensions are "curled up" into circles.

A standard analogy for compactification is to consider a multidimensional object such as a garden hose. If the hose is viewed from a sufficient distance, it appears to have only one dimension, its length, but as one approaches the hose, one discovers that it contains a second dimension, its circumference. Thus, an ant crawling inside it would move in two dimensions.

Quantum Field Theory

The application of quantum mechanics to physical objects such as the electromagnetic field, which are extended in space and time, is known as quantum field theory. In particle physics, quantum field theories form the basis for our understanding of elementary particles, which are modeled as excitations in the fundamental fields. Quantum field theories are also used throughout condensed matter physics to model particle-like objects called quasiparticles.

In the AdS/CFT correspondence, one considers, in addition to a theory of quantum gravity, a certain kind of quantum field theory called a conformal field theory. This is a particularly symmetric and mathematically well behaved type of quantum field theory. Such theories are often studied in the context of string theory, where they are associated with the surface swept out by a string propagating through spacetime, and in statistical mechanics, where they model systems at a thermodynamic critical point.

Overview of the Correspondence

A tessellation of the hyperbolic plane by triangles and squares.

The Geometry of Anti-de Sitter Space

In the AdS/CFT correspondence, one considers string theory or M-theory on an anti-de Sitter background. This means that the geometry of spacetime is described in terms of a certain vacuum solution of Einstein's equation called anti-de Sitter space.

In very elementary terms, anti-de Sitter space is a mathematical model of spacetime in which the notion of distance between points (the metric) is different from the notion of distance in ordinary Euclidean geometry. It is closely related to hyperbolic space, which can be viewed as a disk as illustrated on the right. This image shows a tessellation of a disk by triangles and squares. One can define the distance between points of this disk in such a way that all the triangles and squares are the same size and the circular outer boundary is infinitely far from any point in the interior.

Now imagine a stack of hyperbolic disks where each disk represents the state of the universe at a given time. The resulting geometric object is three-dimensional anti-de Sitter space. It looks like a solid cylinder in which any cross section is a copy of the hyperbolic disk. Time runs along the vertical direction in this picture. The surface of this cylinder plays an important role in the AdS/CFT correspondence. As with the hyperbolic plane, anti-de Sitter space is curved in such a way that any point in the interior is actually infinitely far from this boundary surface.

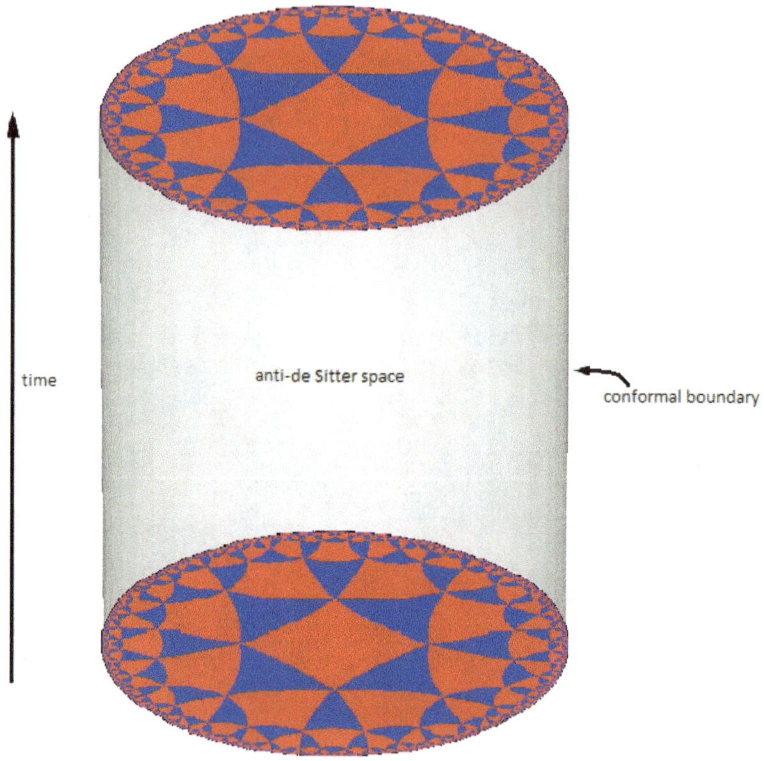

Three-dimensional anti-de Sitter space is like a stack of hyperbolic disks, each one representing the state of the universe at a given time. The resulting spacetime looks like a solid cylinder.

This construction describes a hypothetical universe with only two space and one time dimension, but it can be generalized to any number of dimensions. Indeed, hyperbolic space can have more than two dimensions and one can "stack up" copies of hyperbolic space to get higher-dimensional models of anti-de Sitter space.

The Idea of AdS/CFT

An important feature of anti-de Sitter space is its boundary (which looks like a cylinder in the case of three-dimensional anti-de Sitter space). One property of this boundary is that, locally around any point, it looks just like Minkowski space, the model of spacetime used in nongravitational physics.

One can therefore consider an auxiliary theory in which "spacetime" is given by the boundary of anti-de Sitter space. This observation is the starting point for AdS/CFT correspondence, which states that the boundary of anti-de Sitter space can be regarded as the "spacetime" for a conformal field theory. The claim is that this conformal field theory is equivalent to the gravitational theory on the bulk anti-de Sitter space in the sense that there is a "dictionary" for translating calculations in one theory into calculations in the other. Every entity in one theory has a counterpart in the other theory. For example, a single particle in the gravitational theory might correspond to some collection of particles in the boundary theory. In addition, the predictions in the two theories are quantitatively identical so that if two particles have a 40 percent chance of colliding in the gravitational theory, then the corresponding collections in the boundary theory would also have a 40 percent chance of colliding.

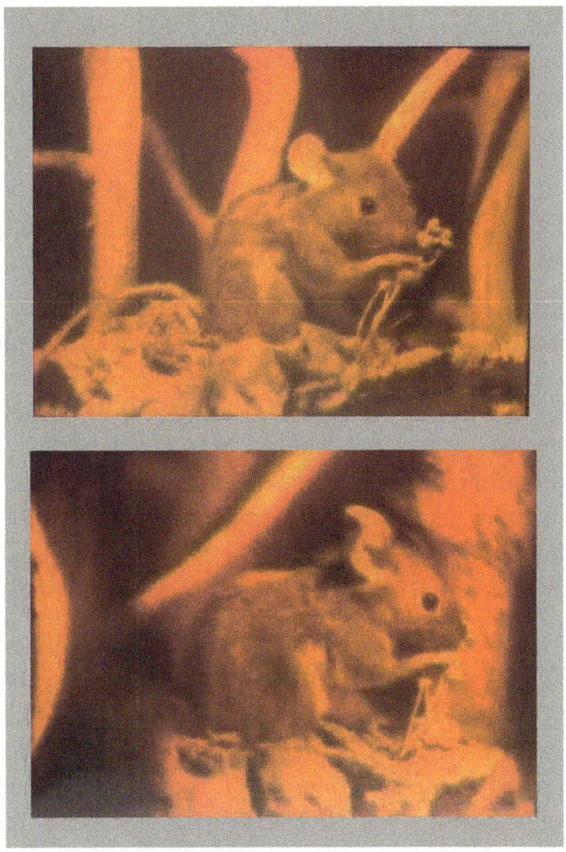

A hologram is a two-dimensional image which stores information about all three dimensions of the object it represents. The two images here are photographs of a single hologram taken from different angles.

Notice that the boundary of anti-de Sitter space has fewer dimensions than anti-de Sitter space itself. For instance, in the three-dimensional example illustrated above, the boundary is a two-dimensional surface. The AdS/CFT correspondence is often described as a "holographic duality" because this relationship between the two theories is similar to the relationship between a three-dimensional object and its image as a hologram. Although a hologram is two-dimensional, it encodes information about all three dimensions of the object it represents. In the same way, theories which are related by the AdS/CFT correspondence are conjectured to be exactly equivalent, despite living in different numbers of dimensions. The conformal field theory is like a hologram which captures information about the higher-dimensional quantum gravity theory.

Examples of the Correspondence

Following Maldacena's insight in 1997, theorists have discovered many different realizations of the AdS/CFT correspondence. These relate various conformal field theories to compactifications of string theory and M-theory in various numbers of dimensions. The theories involved are generally not viable models of the real world, but they have certain features, such as their particle content or high degree of symmetry, which make them useful for solving problems in quantum field theory and quantum gravity.

The most famous example of the AdS/CFT correspondence states that type IIB string theory on the product space $AdS_5 \times S^5$ is equivalent to N = 4 supersymmetric Yang–Mills theory on the four-dimensional boundary. In this example, the spacetime on which the gravitational theory lives is effectively five-dimensional (hence the notation AdS_5) and there are five additional compact dimensions (encoded by the S^5 factor). In the real world, spacetime is four-dimensional, at least macroscopically, so this version of the correspondence does not provide a realistic model of gravity. Likewise, the dual theory is not a viable model of any real-world system as it assumes a large amount of supersymmetry. Nevertheless, as explained below, this boundary theory shares some features in common with quantum chromodynamics, the fundamental theory of the strong force. It describes particles similar to the gluons of quantum chromodynamics together with certain fermions. As a result, it has found applications in nuclear physics, particularly in the study of the quark–gluon plasma.

Another realization of the correspondence states that M-theory on $AdS_7 \times S^4$ is equivalent to the so-called (2,0)-theory in six dimensions. In this example, the spacetime of the gravitational theory is effectively seven-dimensional. The existence of the (2,0)-theory that appears on one side of the duality is predicted by the classification of superconformal field theories. It is still poorly understood because it is a quantum mechanical theory without a classical limit. Despite the inherent difficulty in studying this theory, it is considered to be an interesting object for a variety of reasons, both physical and mathematical.

Yet another realization of the correspondence states that M-theory on $AdS_4 \times S^7$ is equivalent to the ABJM superconformal field theory in three dimensions. Here the gravitational theory has four noncompact dimensions, so this version of the correspondence provides a somewhat more realistic description of gravity.

Applications to Quantum Gravity

A Non-Perturbative Formulation of String Theory

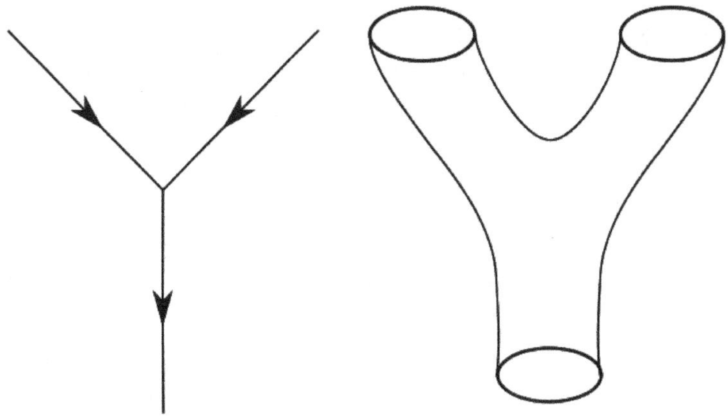

Interaction in the quantum world: world lines of point-like particles or a world sheet swept up by closed strings in string theory.

In quantum field theory, one typically computes the probabilities of various physical events using the techniques of perturbation theory. Developed by Richard Feynman and others in the first half of the twentieth century, perturbative quantum field theory uses special diagrams called Feynman diagrams to organize computations. One imagines that these diagrams depict the paths of point-like particles and their interactions. Although this formalism is extremely useful for making predictions, these predictions are only possible when the strength of the interactions, the coupling constant, is small enough to reliably describe the theory as being close to a theory without interactions.

The starting point for string theory is the idea that the point-like particles of quantum field theory can also be modeled as one-dimensional objects called strings. The interaction of strings is most straightforwardly defined by generalizing the perturbation theory used in ordinary quantum field theory. At the level of Feynman diagrams, this means replacing the one-dimensional diagram representing the path of a point particle by a two-dimensional surface representing the motion of a string. Unlike in quantum field theory, string theory does not yet have a full non-perturbative definition, so many of the theoretical questions that physicists would like to answer remain out of reach.

The problem of developing a non-perturbative formulation of string theory was one of the original motivations for studying the AdS/CFT correspondence. As explained above, the correspondence provides several examples of quantum field theories which are equivalent to string theory on anti-de Sitter space. One can alternatively view this correspondence as providing a definition of string theory in the special case where the gravitational field is asymptotically anti-de Sitter (that is, when the gravitational field resembles that of anti-de Sitter space at spatial infinity). Physically interesting quantities in string theory are defined in terms of quantities in the dual quantum field theory.

Black Hole Information Paradox

In 1975, Stephen Hawking published a calculation which suggested that black holes are not completely black but emit a dim radiation due to quantum effects near the event horizon. At first, Hawking's result posed a problem for theorists because it suggested that black holes destroy information. More precisely, Hawking's calculation seemed to conflict with one of the basic postulates of quantum mechanics, which states that physical systems evolve in time according to the Schrödinger equation. This property is usually referred to as unitarity of time evolution. The apparent contradiction between Hawking's calculation and the unitarity postulate of quantum mechanics came to be known as the black hole information paradox.

The AdS/CFT correspondence resolves the black hole information paradox, at least to some extent, because it shows how a black hole can evolve in a manner consistent with quantum mechanics in some contexts. Indeed, one can consider black holes in the context of the AdS/CFT correspondence, and any such black hole corresponds to a configuration of particles on the boundary of anti-de Sitter space. These particles obey the usual rules of quantum mechanics and in particular evolve in a unitary fashion, so the black hole must also evolve in a unitary fashion, respecting the principles of quantum mechanics. In 2005, Hawking announced that the paradox had been settled in favor of information conservation by the AdS/CFT correspondence, and he suggested a concrete mechanism by which black holes might preserve information.

Applications to Quantum Field Theory

Nuclear Physics

One physical system which has been studied using the AdS/CFT correspondence is the quark–gluon plasma, an exotic state of matter produced in particle accelerators. This state of matter arises for brief instants when heavy ions such as gold or lead nuclei are collided at high energies. Such collisions cause the quarks that make up atomic nuclei to deconfine at temperatures of approximately two trillion kelvins, conditions similar to those present at around 10^{-11} seconds after the Big Bang.

The physics of the quark–gluon plasma is governed by quantum chromodynamics, but this theory is mathematically intractable in problems involving the quark–gluon plasma. In an article appearing in 2005, Đàm Thanh Sơn and his collaborators showed that the AdS/CFT correspondence could be used to understand some aspects of the quark–gluon plasma by describing it in the language of string theory. By applying the AdS/CFT correspondence, Sơn and his collaborators were able to describe the quark gluon plasma in terms of black holes in five-dimensional spacetime. The calculation showed that the ratio of two quantities associated with the quark–gluon plasma, the shear viscosity η and volume density of entropy s, should be approximately equal to a certain universal constant:

$$\frac{\eta}{s} \approx \frac{\hbar}{4\pi k}$$

where $\hbar$ denotes the reduced Planck's constant and k is Boltzmann's constant. In addition, the authors conjectured that this universal constant provides a lower bound for η/s in a large class of systems. In 2008, the predicted value of this ratio for the quark–gluon plasma was confirmed at the Relativistic Heavy Ion Collider at Brookhaven National Laboratory.

Another important property of the quark–gluon plasma is that very high energy quarks moving through the plasma are stopped or "quenched" after traveling only a few femtometers. This phenomenon is characterized by a number $\hat{q}$ called the jet quenching parameter, which relates the energy loss of such a quark to the squared distance traveled through the plasma. Calculations based on the AdS/CFT correspondence have allowed theorists to estimate $\hat{q}$ and the results agree roughly with the measured value of this parameter, suggesting that the AdS/CFT correspondence will be useful for developing a deeper understanding of this phenomenon.

Condensed Matter Physics

A magnet levitating above a high-temperature superconductor. Today some physicists are working to understand high-temperature superconductivity using the AdS/CFT correspondence.

Over the decades, experimental condensed matter physicists have discovered a number of exotic states of matter, including superconductors and superfluids. These states are described using the formalism of quantum field theory, but some phenomena are difficult to explain using standard field theoretic techniques. Some condensed matter theorists including Subir Sachdev hope that the AdS/CFT correspondence will make it possible to describe these systems in the language of string theory and learn more about their behavior.

So far some success has been achieved in using string theory methods to describe the transition of a superfluid to an insulator. A superfluid is a system of electrically neutral atoms that flows without any friction. Such systems are often produced in the laboratory using liquid helium, but recently experimentalists have developed new ways of producing artificial superfluids by pouring trillions of cold atoms into a lattice of criss-crossing lasers. These atoms initially behave as a superfluid, but as experimentalists increase the intensity of the lasers, they become less mobile and then suddenly transition to an insulating state. During the transition, the atoms behave in an unusual way. For example, the atoms slow to a halt at a rate that depends on the temperature and on Planck's constant, the fundamental parameter of quantum mechanics, which does not enter into the description of the other phases. This behavior has recently been understood by considering a dual description where properties of the fluid are described in terms of a higher dimensional black hole.

Criticism

With many physicists turning towards string-based methods to attack problems in nuclear and condensed matter physics, some theorists working in these areas have expressed doubts about whether the AdS/CFT correspondence can provide the tools needed to realistically model real-world systems. In a talk at the Quark Matter conference in 2006, Larry McLerran pointed out that the N=4 super Yang–Mills theory that appears in the AdS/CFT correspondence differs significantly from quantum chromodynamics, making it difficult to apply these methods to nuclear physics. According to McLerran,

N=4 supersymmetric Yang–Mills is not QCD ... It has no mass scale and is conformally invariant. It has no confinement and no running coupling constant. It is supersymmetric. It has no chiral symmetry breaking or mass generation. It has six scalar and fermions in the adjoint representation ... It may be possible to correct some or all of the above problems, or, for various physical problems, some of the objections may not be relevant. As yet there is not consensus nor compelling arguments for the conjectured fixes or phenomena which would insure that the N=4 supersymmetric Yang Mills results would reliably reflect QCD.

In a letter to Physics Today, Nobel laureate Philip W. Anderson voiced similar concerns about applications of AdS/CFT to condensed matter physics, stating

As a very general problem with the AdS/CFT approach in condensed-matter theory, we can point to those telltale initials CFT—conformal field theory. Condensed-matter problems are, in general, neither relativistic nor conformal. Near a quantum critical point, both time and space may be scaling, but even there we still have a preferred coordinate system and, usually, a lattice. There is some evidence of other linear-T phases to the left of the strange metal about which they are welcome to speculate, but again in this case the condensed-matter problem is overdetermined by experimental facts.

Gerard ‘t Hooft obtained results related to the AdS/CFT correspondence in the 1970s by studying analogies between string theory and nuclear physics.

String Theory and Nuclear Physics

The discovery of the AdS/CFT correspondence in late 1997 was the culmination of a long history of efforts to relate string theory to nuclear physics. In fact, string theory was originally developed during the late 1960s and early 1970s as a theory of hadrons, the subatomic particles like the proton and neutron that are held together by the strong nuclear force. The idea was that each of these particles could be viewed as a different oscillation mode of a string. In the late 1960s, experimentalists had found that hadrons fall into families called Regge trajectories with squared energy proportional to angular momentum, and theorists showed that this relationship emerges naturally from the physics of a rotating relativistic string.

On the other hand, attempts to model hadrons as strings faced serious problems. One problem was that string theory includes a massless spin-2 particle whereas no such particle appears in the physics of hadrons. Such a particle would mediate a force with the properties of gravity. In 1974, Joel Scherk and John Schwarz suggested that string theory was therefore not a theory of nuclear physics as many theorists had thought but instead a theory of quantum gravity. At the same time, it was realized that hadrons are actually made of quarks, and the string theory approach was abandoned in favor of quantum chromodynamics.

In quantum chromodynamics, quarks have a kind of charge that comes in three varieties called colors. In a paper from 1974, Gerard 't Hooft studied the relationship between string theory and nuclear physics from another point of view by considering theories similar to quantum chromodynamics, where the number of colors is some arbitrary number N, rather than three. In this article, 't Hooft considered a certain limit where N tends to infinity and argued that in this limit certain calculations in quantum field theory resemble calculations in string theory.

Stephen Hawking predicted in 1975 that black holes emit radiation due to quantum effects.

Black Holes and Holography

In 1975, Stephen Hawking published a calculation which suggested that black holes are not completely black but emit a dim radiation due to quantum effects near the event horizon. This work extended previous results of Jacob Bekenstein who had suggested that black holes have a well defined entropy. At first, Hawking's result appeared to contradict one of the main postulates of quantum mechanics, namely the unitarity of time evolution. Intuitively, the unitarity postulate says that quantum mechanical systems do not destroy information as they evolve from one state to another. For this reason, the apparent contradiction came to be known as the black hole information paradox.

Later, in 1993, Gerard 't Hooft wrote a speculative paper on quantum gravity in which he revisited Hawking's work on black hole thermodynamics, concluding that the total number of degrees of freedom in a region of spacetime surrounding a black hole is proportional to the surface area of the horizon. This idea was promoted by Leonard Susskind and is now known as the holographic principle. The holographic principle and its realization in string theory through the AdS/CFT correspondence have helped elucidate the mysteries of black holes suggested by Hawking's work and are believed to provide a resolution of the black hole information paradox. In 2004, Hawking conceded that black holes do not violate quantum mechanics, and he suggested a concrete mechanism by which they might preserve information.

Leonard Susskind made early contributions to the idea of holography in quantum gravity.

Maldacena's Paper

In late 1997, Juan Maldacena published a landmark paper that initiated the study of AdS/CFT. According to Alexander Markovich Polyakov, "Maldacena's work opened the flood gates." The conjecture immediately excited great interest in the string theory community and was considered in articles by Steven Gubser, Igor Klebanov and Polyakov, and by Edward Witten. These papers made Maldacena's conjecture more precise and showed that the conformal field theory appearing in the correspondence lives on the boundary of anti-de Sitter space.

Juan Maldacena first proposed the AdS/CFT correspondence in late 1997.

One special case of Maldacena's proposal says that N=4 super Yang–Mills theory, a gauge theory similar in some ways to quantum chromodynamics, is equivalent to string theory in five-dimensional anti-de Sitter space. This result helped clarify the earlier work of 't Hooft on the relationship between string theory and quantum chromodynamics, taking string theory back to its roots as a theory of nuclear physics. Maldacena's results also provided a concrete realization of the holographic principle with important implications for quantum gravity and black hole physics. By the year 2015, Maldacena's paper had become the most highly cited paper in high energy physics with over 10,000 citations. These subsequent articles have provided considerable evidence that the correspondence is correct, although so far it has not been rigorously proved.

AdS/CFT Finds Applications

In 1999, after taking a job at Columbia University, nuclear physicist Đàm Thanh Sơn paid a visit to Andrei Starinets, a friend from Sơn's undergraduate days who happened to be doing a Ph.D. in string theory at New York University. Although the two men had no intention of collaborating, Sơn soon realized that the AdS/CFT calculations Starinets was doing could shed light on some aspects of the quark–gluon plasma, an exotic state of matter produced when heavy ions are collided at high energies. In collaboration with Starinets and Pavel Kovtun, Sơn was able to use the AdS/CFT correspondence to calculate a key parameter of the plasma. As Sơn later recalled, "We turned the calculation on its head to give us a prediction for the value of the shear viscosity of a plasma ... A friend of mine in nuclear physics joked that ours was the first useful paper to come out of string theory."

Today physicists continue to look for applications of the AdS/CFT correspondence in quantum field theory. In addition to the applications to nuclear physics advocated by Đàm Thanh Sơn and his collaborators, condensed matter physicists such as Subir Sachdev have used string theory methods to understand some aspects of condensed matter physics. A notable result in this direction was the description, via the AdS/CFT correspondence, of the transition of a superfluid to an insulator. Another emerging subject is the fluid/gravity correspondence, which uses the AdS/CFT correspondence to translate problems in fluid dynamics into problems in general relativity.

Generalizations

Three-Dimensional Gravity

In order to better understand the quantum aspects of gravity in our four-dimensional universe, some physicists have considered a lower-dimensional mathematical model in which spacetime has only two spatial dimensions and one time dimension. In this setting, the mathematics describing the gravitational field simplifies drastically, and one can study quantum gravity using familiar methods from quantum field theory, eliminating the need for string theory or other more radical approaches to quantum gravity in four dimensions.

Beginning with the work of J. D. Brown and Marc Henneaux in 1986, physicists have noticed that quantum gravity in a three-dimensional spacetime is closely related to two-dimensional conformal field theory. In 1995, Henneaux and his coworkers explored this relationship in more detail, suggesting that three-dimensional gravity in anti-de Sitter space is equivalent to the conformal field theory known as Liouville field theory.Another conjecture formulated by Edward Witten states that three-dimensional gravity in anti-de Sitter space is equivalent to a conformal field theory with monster group symmetry. These conjectures provide examples of the AdS/CFT correspondence that do not require the full apparatus of string or M-theory.

dS/CFT Correspondence

Unlike our universe, which is now known to be expanding at an accelerating rate, anti-de Sitter space is neither expanding nor contracting. Instead it looks the same at all times. In more technical language, one says that anti-de Sitter space corresponds to a universe with a negative cosmological constant, whereas the real universe has a small positive cosmological constant.

Although the properties of gravity at short distances should be somewhat independent of the value of the cosmological constant, it is desirable to have a version of the AdS/CFT correspondence for positive cosmological constant. In 2001, Andrew Strominger introduced a version of the duality called the dS/CFT correspondence. This duality involves a model of spacetime called de Sitter space with a positive cosmological constant. Such a duality is interesting from the point of view of cosmology since many cosmologists believe that the very early universe was close to being de Sitter space. Our universe may also resemble de Sitter space in the distant future.

Kerr/CFT Correspondence

Although the AdS/CFT correspondence is often useful for studying the properties of black holes, most of the black holes considered in the context of AdS/CFT are physically unrealistic. Indeed, as explained above, most versions of the AdS/CFT correspondence involve higher-dimensional models of spacetime with unphysical supersymmetry.

In 2009, Monica Guica, Thomas Hartman, Wei Song, and Andrew Strominger showed that the ideas of AdS/CFT could nevertheless be used to understand certain astrophysical black holes. More precisely, their results apply to black holes that are approximated by extremal Kerr black holes, which have the largest possible angular momentum compatible with a given mass. They showed that such black holes have an equivalent description in terms of conformal field theory. The Kerr/CFT correspondence was later extended to black holes with lower angular momentum.

Higher Spin Gauge Theories

The AdS/CFT correspondence is closely related to another duality conjectured by Igor Klebanov and Alexander Markovich Polyakov in 2002.This duality states that certain "higher spin gauge theories" on anti-de Sitter space are equivalent to conformal field theories with O(N) symmetry. Here the theory in the bulk is a type of gauge theory describing particles of arbitrarily high spin. It is similar to string theory, where the excited modes of vibrating strings correspond to particles with higher spin, and it may help to better understand the string theoretic versions of AdS/CFT and possibly even prove the correspondence. In 2010, Simone Giombi and Xi Yin obtained further evidence for this duality by computing quantities called three-point functions.

LIST OF CONJECTURES

This is a list of mathematical conjectures.

Open Problems

- 1/3–2/3 conjecture
- abc conjecture (proof claimed by Shinichi Mochizuki)
- Agoh–Giuga conjecture
- Andrews–Curtis conjecture
- Andrica's conjecture
- Artin conjecture (L-functions)
- Artin's conjecture on primitive roots
- Bateman–Horn conjecture
- Baum–Connes conjecture
- Beal's conjecture
- Beilinson conjecture
- Berry–Tabor conjecture
- Birch and Swinnerton-Dyer conjecture
- Birch–Tate conjecture
- Birkhoff conjecture
- Bloch–Beilinson conjectures
- Bloch–Kato conjecture
- Bombieri-Lang conjecture
- Borel conjecture
- Bost conjecture
- Brocard's conjecture
- Brumer–Stark conjecture
- Bunyakovsky conjecture
- Carathéodory conjecture
- Carmichael totient conjecture
- Casas-Alvero conjecture

- Catalan–Dickson conjecture on aliquot sequences
- Catalan's Mersenne conjecture
- Cherlin–Zilber conjecture
- Collatz conjecture
- Cramér's conjecture
- Conway's thrackle conjecture
- Deligne conjecture in monodromy
- Eilenberg−Ganea conjecture
- Elliott–Halberstam conjecture
- Erdős–Burr conjecture
- Erdős–Faber–Lovász conjecture
- Erdős–Gyárfás conjecture
- Erdős–Straus conjecture
- Farrell–Jones conjecture
- Filling area conjecture
- Firoozbakht's conjecture
- Fortune's conjecture
- Frankl conjecture
- Gauss circle problem
- Gilbreath conjecture
- Giuga's conjecture
- Goldbach's conjecture
- Goldbach's weak conjecture (proof claimed by Harald Helfgott)
- Gold partition conjecture
- Goormaghtigh conjecture
- Green's conjecture
- Grimm's conjecture
- Grothendieck–Katz p-curvature conjecture
- Guralnick–Thompson conjecture (Riemann surfaces)
- Hadamard conjecture
- Hedetniemi's conjecture
- Herzog–Schönheim conjecture
- Hilbert–Smith conjecture
- Hodge conjecture
- Homological conjectures in commutative algebra
- Hopf conjecture

- Invariant subspace problem
- Jacobian conjecture
- Jacobson's conjecture
- Kakeya conjectures
- Kakeya needle problem
- Kaplansky conjecture
- Keating–Snaith conjecture
- Köthe conjecture
- Kung-Traub conjecture
- Lawson's conjecture
- Lemoine's conjecture
- Lenstra–Pomerance–Wagstaff conjecture
- Leopoldt's conjecture
- Lichtenbaum conjecture
- List coloring conjecture
- Littlewood conjecture
- Lovász conjecture
- MNOP conjecture
- Manin conjecture
- Marshall Hall's conjecture
- Mazur's conjectures
- Montgomery's pair correlation conjecture
- n conjecture
- Nagata's conjecture on automorphisms
- New Mersenne conjecture
- Nirenberg–Treves conjecture
- Novikov conjecture
- Oppermann's conjecture
- Petersen coloring conjecture
- Pierce–Birkhoff conjecture
- Pillai's conjecture
- De Polignac's conjecture
- quantum unique ergodicity conjecture, (2004, Elon Lindenstrauss, for arithmetic hyperbolic surfaces, 2008, Kannan Soundararajan & Roman Holowinsky, for holomorphic forms of increasing weight for Hecke eigenforms on noncompact arithmetic surfaces)
- Quillen–Lichtenbaum conjecture

- Reconstruction conjecture
- Riemann Hypotheses: see also Weil conjectures under Theorems
- Riemann hypothesis
- Generalized Riemann hypothesis
- Grand Riemann hypothesis
- Density hypothesis
- Lindelöf hypothesis
- Hilbert–Pólya conjecture on the Riemann hypothesis
- Ringel-Kotzig conjecture
- Sato–Tate conjecture
- Schanuel's conjecture
- Schinzel's hypothesis H
- Scholz conjecture
- Second Hardy–Littlewood conjecture
- Selfridge's conjecture
- Sendov's conjecture
- Serre's multiplicity conjectures
- Singmaster's conjecture
- Standard conjectures on algebraic cycles
- Tate conjecture
- Toeplitz' conjecture
- Twin prime conjecture
- Ulam's packing conjecture
- Unicity conjecture for Markov numbers
- Uniformity conjecture
- Vandiver's conjecture
- Vizing's conjecture
- Waring's conjecture
- Weight-monodromy conjecture
- Weinstein conjecture
- Whitehead conjecture
- Zauner's conjecture
- Zhou conjecture

Proved (Now Theorems)

The theorems may not be their official names.

- Adams conjecture (J-homomorphism)
- Andrews–Curtis conjecture
- Angel problem (several proofs, 2007)
- Bieberbach conjecture, 1916 (De Branges' theorem, 1984)
- Blattner's conjecture (Blattner formula)
- Bounded gap conjecture (Zhang Yitang, 2013)
- Burnside conjecture (Feit–Thompson theorem)
- Cameron–Erdős conjecture
- Catalan's conjecture, 1844 (Mihăilescu's theorem, 2002)
- Circular law (proved in 2010)
- Conway–Norton conjecture (monstrous moonshine)
- Dinitz conjecture (Galvin's theorem)
- Deligne's conjecture on 1-motives
- Denjoy's conjecture
- Dodecahedral conjecture (Hales–McLaughlin theorem)
- Ehrenpreis conjecture (proved in 2011)
- Epsilon conjecture (Ribet's theorem)
- Fermat's Last Theorem (Andrew Wiles' proof in 1994)
- Four color theorem (Appel and Haken, 1976)
- Frobenius conjecture (Iiyori and Yamaki)
- Gradient conjecture (Kurdyka–Mostowski–Parusinski theorem)
- Hanna Neumann conjecture (proved in 2011)
- Heawood conjecture (Ringel–Youngs' theorem)
- Kadison–Singer problem (Marcus-Spielman-Srivastava theorem)
- Kepler conjecture (proved in 1998)
- Kummer's conjecture on cubic Gauss sums (Kummer sum)
- Mahler–Manin conjecture
- Main conjecture in Vinogradov's mean-value theorem (Bourgain–Demeter–Guth theorem)
- Manin–Mumford conjecture (Raynaud's theorem)
- Milnor conjecture (Voevodsky's theorem)
- Mordell conjecture (Faltings' theorem)
- Mumford conjecture (Haboush's theorem)
- n! conjecture (proved in 2001)
- Poincaré conjecture, 1904 (theorem about the characterization of the three-dimensional sphere)
- Oppenheim conjecture (Margulis' theorem)

- Ramanujan–Petersson conjecture (consequence of the proof of the Weil conjectures)
- Road coloring conjecture, 1970 (2008)
- Scheinerman's conjecture (proved 2009)
- Segal's Burnside ring conjecture (Carlsson theorem)
- Serre's conjecture (Quillen–Suslin theorem)
- Serre's modularity conjecture (proved 2008)
- Seymour's conjecture
- Smith conjecture (Gordon's theorem)
- Stanley–Wilf conjecture (Marcus–Tardos theorem)
- Star height problem (Hashiguchi theorem)
- Strong perfect graph conjecture (Chudnovsky–Robertson–Seymour–Thomas theorem)
- Sullivan conjecture
- Surface subgroup conjecture (proved in 2009)
- Tameness conjecture (Agol or Calegari–Gabai theorem)
- Taniyama–Shimura conjecture (Modularity theorem)
- Thurston's geometrization conjecture (3-manifolds of the uniformization theorem for surfaces)
- Wagner's conjecture (Robertson–Seymour theorem)
- Weil conjectures (Deligne's theorems)
- Willmore conjecture (proved in 2012)

Disproved (No Longer Conjectures)

- Atiyah conjecture (not a conjecture to start with)
- Borsuk's conjecture
- Chinese hypothesis (not a conjecture to start with)
- Euler's sum of powers conjecture
- Ganea conjecture
- Generalized Smith conjecture
- Hauptvermutung
- Hirsch conjecture (disproved in 2010)
- Intersection graph conjecture
- Kouchnirenko's conjecture
- Mertens conjecture
- Pólya conjecture, 1919 (1958)

- Ragsdale conjecture
- Tait's conjecture
- Von Neumann conjecture
- Weyl–Berry conjecture

REFERENCES

1. Oxford Dictionary of English (2010 ed.).
2. Schwartz, JL (1995). Shuttling between the particular and the general: reflections on the role of conjecture and hypothesis in the generation of knowledge in science and mathematics. p. 93.
3. Ore, Oystein (1988) , Number Theory and Its History, Dover, pp. 203–204, ISBN 978-0-486-65620-5
4. Georges Gonthier (December 2008). "Formal Proof—The Four-Color Theorem". Notices of the AMS. 55 (11): 1382–1393.From this paper: Definitions: A planar map is a set of pairwise disjoint subsets of the plane, called regions. A simple map is one whose regions are connected open sets. Two regions of a map are adjacent if their respective closures have a common point that is not a corner of the map. A point is a corner of a map if and only if it belongs to the closures of at least three regions. Theorem: The regions of any simple planar map can be colored with only four colors, in such a way that any two adjacent regions have different colors.
5. W. W. Rouse Ball (1960) The Four Color Theorem, in Mathematical Recreations and Essays, Macmillan, New York, pp 222-232.
6. Donald MacKenzie, Mechanizing Proof: Computing, Risk, and Trust (MIT Press, 2004) p103
7. Milnor, John W. (1961). "Two complexes which are homeomorphic but combinatorially distinct". Annals of Mathematics. 74 (2): 575–590. doi:10.2307/1970299. JSTOR 1970299. MR 133127.
8. Moise, Edwin E. (1977). Geometric Topology in Dimensions 2 and 3. New York: New York : Springer-Verlag. ISBN 978-0-387-90220-3.
9. Hamilton, Richard S. (1997). "Four-manifolds with positive isotropic curvature". Communications in Analysis and Geometry. 5 (1): 1–92. MR 1456308. Zbl 0892.53018.
10. Juris Hartmanis 1989, Gödel, von Neumann, and the P = NP problem, Bulletin of the European Association for Theoretical Computer Science, vol. 38, pp. 101–107
11. Cook, Stephen (1971). "The complexity of theorem proving procedures". Proceedings of the Third Annual ACM Symposium on Theory of Computing. pp. 151–158.
12. Lance Fortnow, The status of the P versus NP problem, Communications of the ACM 52 (2009), no. 9, pp. 78–86. doi:10.1145/1562164.1562186

13. Langlands, Robert (1967), Letter to Prof. Weil
14. Popper, Karl (2004). Conjectures and refutations : the growth of scientific knowledge. London: Routledge. ISBN 0-415-28594-1.
15. Singh, pp. 18–20.
16. "Science and Technology". The Guinness Book of World Records. Guinness Publishing Ltd. 1995.
17. "Fermat's last theorem earns Andrew Wiles the Abel Prize". Nature. 15 March 2016. Retrieved 15 March 2016.
18. British mathematician Sir Andrew Wiles gets Abel math prize - The Washington Post
19. 300-year-old math question solved, professor wins $700k - CNN.com
20. Wiles, Andrew (1995). "Modular elliptic curves and Fermat's Last Theorem" (PDF). Annals of Mathematics. 141 (3): 448. doi:10.2307/2118559. JSTOR 2118559. OCLC 37032255. Frey's suggestion, in the notation of the following theorem, was to show that the (hypothetical) elliptic curve y2 = x(x + up)(x - vp) could not be modular.
21. Ribet, Ken (1990). "On modular representations of Gal(Q/Q) arising from modular forms" (PDF). Inventiones mathematicae. 100 (2): 432. doi:10.1007/BF01231195. MR 1047143.
22. Stark, pp. 151–155.
23. Stillwell J (2003). Elements of Number Theory. New York: Springer-Verlag. pp. 110–112. ISBN 0-387-95587-9. Retrieved 2016-03-17.
24. Aczel, pp. 13–15
25. Singh, p. 6.
26. Stark, pp. 145–146.
27. Singh, pp. 50–51.
28. Stark, p. 145.
29. Aczel, pp. 44–45; Singh, pp. 56–58.
30. Aczel, pp. 14–15.
31. Stark, pp. 44–47.
32. Friberg, pp. 333– 334.
33. Dickson, p. 731; Singh, pp. 60–62; Aczel, p. 9.
34. T. Heath, Diophantus of Alexandria Second Edition, Cambridge University Press, 1910, reprinted by Dover, NY, 1964, pp. 144-145
35. Panchishkin, p. 341
36. Singh, pp. 62–66.
37. Dickson, p. 731.
38. Singh, p. 67; Aczel, p. 10.
39. Ribenboim, pp. 13, 24.

40. van der Poorten, Notes and Remarks 1.2, p. 5.
41. van der Poorten, loc. cit.
42. André Weil (1984). Number Theory: An approach through history. From Hammurapi to Legendre. Basel, Switzerland: Birkhäuser. p. 104.
43. BBC Documentary.
44. Freeman L. "Fermat's One Proof". Retrieved 23 May 2009.
45. Dickson, pp. 615–616; Aczel, p. 44.
46. Ribenboim, pp. 15–24.
47. Frénicle de Bessy, Traité des Triangles Rectangles en Nombres, vol. I, 1676, Paris. Reprinted in Mém. Acad. Roy. Sci., 5, 1666–1699 (1729).
48. Euler L (1738). "Theorematum quorundam arithmeticorum demonstrationes". Comm. Acad. Sci. Petrop. 10: 125–146.. Reprinted Opera omnia, ser. I, "Commentationes Arithmeticae", vol. I, pp. 38–58, Leipzig:Teubner (1915).
49. Kausler CF (1802). "Nova demonstratio theorematis nec summam, nec differentiam duorum cuborum cubum esse posse". Novi Acta Acad. Petrop. 13: 245–253.
50. Barlow P (1811). An Elementary Investigation of Theory of Numbers. St. Paul's Church-Yard, London: J. Johnson. pp. 144–145.
51. Legendre AM (1830). Théorie des Nombres (Volume II) (3rd ed.). Paris: Firmin Didot Frères. Reprinted in 1955 by A. Blanchard (Paris).
52. Schopis (1825). Einige Sätze aus der unbestimmten Analytik. Gummbinnen: Programm.
53. Terquem O (1846). "Théorèmes sur les puissances des nombres". Nouv. Ann. Math. 5: 70–87.
54. Bertrand J (1851). Traité Élémentaire d'Algèbre. Paris: Hachette. pp. 217–230, 395.
55. Lebesgue VA (1853). "Résolution des équations biquadratiques z2 = x4 ± 2my4, z2 = 2mx4 − y4, 2mz2 = x4 ± y4". J. Math. Pures Appl. 18: 73–86.
56. Pepin T (1883). "Étude sur l'équation indéterminée ax4 + by4 = cz2". Atti Accad. Naz. Lincei. 36: 34–70.
57. Tafelmacher WLA (1893). "Sobre la ecuación x4 + y4 = z4". Ann. Univ. Chile. 84: 307–320.
58. Hilbert D (1897). "Die Theorie der algebraischen Zahlkörper". Jahresbericht der Deutschen Mathematiker-Vereinigung. 4: 175–546.Reprinted in 1965 in Gesammelte Abhandlungen, vol. I by New York:Chelsea.
59. Bendz TR (1901). Öfver diophantiska ekvationen xn + yn = zn. Uppsala: Almqvist & Wiksells Boktrycken.
60. Gambioli D (1901). "Memoria bibliographica sull'ultimo teorema di Fermat". Period. Mat. 16: 145–192.
61. Kronecker L (1901). Vorlesungen über Zahlentheorie, vol. I. Leipzig: Teubner. pp. 35–38. Reprinted by New York:Springer-Verlag in 1978.

62. Bang A (1905). "Nyt Bevis for at Ligningen x4 − y4 = z4, ikke kan have rationale Løsinger". Nyt Tidsskrift Mat. 16B: 35–36.
63. Sommer J (1907). Vorlesungen über Zahlentheorie. Leipzig: Teubner.
64. Bottari A (1908). "Soluzione intere dell'equazione pitagorica e applicazione alla dimostrazione di alcune teoremi della teoria dei numeri". Period. Mat. 23: 104–110.
65. Rychlik K (1910). "On Fermat's last theorem for n = 4 and n = 3 (in Bohemian)". Časopis Pěst. Mat. 39: 65–86.
66. Nutzhorn F (1912). "Den ubestemte Ligning x4 + y4 = z4". Nyt Tidsskrift Mat. 23B: 33–38.
67. Carmichael RD (1913). "On the impossibility of certain Diophantine equations and systems of equations". Amer. Math. Monthly. Mathematical Association of America. 20 (7): 213–221. doi:10.2307/2974106. JSTOR 2974106.
68. Hancock H (1931). Foundations of the Theory of Algebraic Numbers, vol. I. New York: Macmillan.
69. Vrănceanu G (1966). "Asupra teorema lui Fermat pentru n=4". Gaz. Mat. Ser. A. 71: 334–335. Reprinted in 1977 in Opera matematica, vol. 4, pp. 202–205, Bucureşti:Edit. Acad. Rep. Soc. Romana.
70. Grant, Mike, and Perella, Malcolm, "Descending to the irrational", Mathematical Gazette 83, July 1999, pp.263–267.
71. Barbara, Roy, "Fermat's last theorem in the case n=4", Mathematical Gazette 91, July 2007, 260–262.
72. Dolan, Stan, "Fermat's method of descente infinie", Mathematical Gazette 95, July 2011, 269–271.
73. Ribenboim, pp. 1–2.
74. Dickson, p. 545.
75. Euler L (1770) Vollständige Anleitung zur Algebra, Roy. Acad. Sci., St. Petersburg.
76. Freeman L. "Fermat's Last Theorem: Proof for n = 3". Retrieved 23 May 2009.
77. Ribenboim, pp. 24–25; Mordell, pp. 6–8; Edwards, pp. 39–40.
78. Aczel, p. 44; Edwards, pp. 40, 52–54.
79. Ribenboim, pp. 33, 37–41.
80. Legendre AM (1823). "Recherches sur quelques objets d'analyse indéterminée, et particulièrement sur le théorème de Fermat". Mém. Acad. Roy. Sci. Institut France. 6: 1–60. Reprinted in 1825 as the "Second Supplément" for a printing of the 2nd edition of Essai sur la Théorie des Nombres, Courcier (Paris). Also reprinted in 1909 in Sphinx-Oedipe, 4, 97–128.
81. Calzolari L (1855). Tentativo per dimostrare il teorema di Fermat sull'equazione indeterminata xn + yn = zn. Ferrara.
82. Lamé G (1865). "Étude des binômes cubiques x3 ± y3". C. R. Acad. Sci. Paris. 61: 921–924, 961–965.

83. Tait PG (1872). “Mathematical Notes”. Proc. Roy. Soc. Edinburgh. 7: 144.
84. Günther S (1878). “Über die unbestimmte Gleichung x3 + y3 = z3”. Sitzungsberichte Böhm. Ges. Wiss.: 112–120.
85. Krey H (1909). “Neuer Beweis eines arithmetischen Satzes”. Math. Naturwiss. Blätter. 6: 179–180.
86. Stockhaus H (1910). Beitrag zum Beweis des Fermatschen Satzes. Leipzig: Brandstetter.
87. Carmichael RD (1915). Diophantine Analysis. New York: Wiley.
88. van der Corput JG (1915). “Quelques formes quadratiques et quelques équations indéterminées”. Nieuw Archief Wisk. 11: 45–75.
89. Thue A (1917). “Et bevis for at ligningen A3 + B3 = C3 er unmulig i hele tal fra nul forskjellige tal A, B og C”. Arch. Mat. Naturv. 34 (15).Reprinted in Selected Mathematical Papers (1977), Oslo:Universitetsforlaget, pp. 555–559.
90. Duarte FJ (1944). “Sobre la ecuación x3 + y3 + z3 = 0”. Ciencias Fis. Mat. Naturales (Caracas). 8: 971–979.
91. Freeman L. “Fermat’s Last Theorem: Proof for n = 5”. Retrieved 23 May 2009.
92. Ribenboim, p. 49; Mordell, p. 8–9; Aczel, p. 44; Singh, p. 106.
93. Ribenboim, pp. 55–57.
94. Gauss CF (1875). “Neue Theorie der Zerlegung der Cuben”. Zur Theorie der complexen Zahlen, Werke, vol. II (2nd ed.). Königl. Ges. Wiss. Göttingen. pp. 387–391. (Published posthumously)
95. Lebesgue VA (1843). “Théorèmes nouveaux sur l’équation indéterminée x5 + y5 = az5”. J. Math. Pures Appl. 8: 49–70.
96. Lamé G (1847). “Mémoire sur la résolution en nombres complexes de l’équation A5 + B5 + C5 = 0”. J. Math. Pures Appl. 12: 137–171.
97. Gambioli D (1903–1904). “Intorno all’ultimo teorema di Fermat”. Il Pitagora. 10: 11–13, 41–42.
98. Werebrusow AS (1905). “On the equation x5 + y5 = Az5 (in Russian)”. Moskov. Math. Samml. 25: 466–473.
99. Rychlik K (1910). “On Fermat’s last theorem for n = 5 (in Bohemian)”. Časopis Pěst. Mat. 39: 185–195, 305–317.
100. Terjanian G (1987). “Sur une question de V. A. Lebesgue”. Annales de l’Institut Fourier. 37: 19–37. doi:10.5802/aif.1096.
101. Ribenboim, pp. 57–63; Mordell, p. 8; Aczel, p. 44; Singh, p. 106.
102. Lamé G (1839). “Mémoire sur le dernier théorème de Fermat”. C. R. Acad. Sci. Paris. 9: 45–46.
103. Lebesgue VA (1840). “Démonstration de l’impossibilité de résoudre l’équation x7 + y7 + z7 = 0 en nombres entiers”. J. Math. Pures Appl. 5: 276–279, 348–349.
104. Freeman L. “Fermat’s Last Theorem: Proof for n = 7”. Retrieved 23 May 2009.
105. Genocchi A (1864). “Intorno all’equazioni x7 + y7 + z7 = 0”. Annali Mat. 6:

287–288.

106. Pepin T (1876). "Impossibilité de l'équation x7 + y7 + z7 = 0". C. R. Acad. Sci. Paris. 82: 676–679, 743–747.
107. Maillet E (1897). "Sur l'équation indéterminée axλt + byλt = czλt". Assoc. Française Avanc. Sci., St. Etienne (sér. II). 26: 156–168.
108. Thue A (1896). "Über die Auflösbarkeit einiger unbestimmter Gleichungen". Det Kongel. Norske Videnskabers Selskabs Skrifter. 7.Reprinted in Selected Mathematical Papers, pp. 19–30, Oslo:Universitetsforlaget (1977).
109. Tafelmacher WLA (1897). "La ecuación x3 + y3 = z2: Una demonstración nueva del teorema de fermat para el caso de las sestas potencias". Ann. Univ. Chile, Santiago. 97: 63–80.
110. Lind B (1909). "Einige zahlentheoretische Sätze". Arch. Math. Phys. 15: 368–369.
111. Kapferer H (1913). "Beweis des Fermatschen Satzes für die Exponenten 6 und 10". Arch. Math. Phys. 21: 143–146.
112. Swift E (1914). "Solution to Problem 206". Amer. Math. Monthly. 21: 238–239.
113. Breusch R (1960). "A simple proof of Fermat's last theorem for n = 6, n = 10". Math. Mag. 33 (5): 279–281. doi:10.2307/3029800. JSTOR 3029800.
114. Dirichlet PGL (1832). "Démonstration du théorème de Fermat pour le cas des 14e puissances". J. Reine Angew. Math. 9: 390–393. Reprinted in Werke, vol. I, pp. 189–194, Berlin:G. Reimer (1889); reprinted New York:Chelsea (1969).
115. Terjanian G (1974). "L'équation x14 + y14 = z14 en nombres entiers". Bull. Sci. Math. (sér. 2). 98: 91–95.
116. Edwards, pp. 73–74.
117. Edwards, p. 74.
118. Dickson, p. 733.
119. Ribenboim P (1979). 13 Lectures on Fermat's Last Theorem. New York: Springer Verlag. pp. 51–54. ISBN 978-0-387-90432-0.
120. Singh, pp. 97–109.
121. Laubenbacher R, Pengelley D (2007). "Voici ce que j'ai trouvé: Sophie Germain's grand plan to prove Fermat's Last Theorem"(PDF). Retrieved 19 May 2009.
122. Aczel, p. 57.
123. Terjanian, G. (1977). "Sur l'équation x2p + y2p = z2p". Comptes rendus hebdomadaires des séances de l'Académie des sciences. Série a et B. 285: 973–975.
124. Adleman LM, Heath-Brown DR (June 1985). "The first case of Fermat's last theorem". Inventiones Mathematicae. Berlin: Springer. 79 (2): 409–416. doi:10.1007/BF01388981.
125. Aczel, pp. 84–88; Singh, pp. 232–234.
126. Faltings G (1983). "Endlichkeitssätze für abelsche Varietäten über Zahlkörpern". Inventiones Mathematicae. 73 (3): 349–366. doi:10.1007/BF01388432.

127. Ribenboim P (1979). 13 Lectures on Fermat's Last Theorem. New York: Springer Verlag. p. 202. ISBN 978-0-387-90432-0.
128. Wagstaff SS, Jr. (1978). "The irregular primes to 125000". Math. Comp. American Mathematical Society. 32 (142): 583–591. doi:10.2307/2006167. JSTOR 2006167. (PDF) Archived 10 January 2011 at WebCite
129. Buhler J, Crandell R, Ernvall R, Metsänkylä T (1993). "Irregular primes and cyclotomic invariants to four million". Math. Comp. American Mathematical Society. 61 (203): 151–153. doi:10.2307/2152942. JSTOR 2152942.
130. Fermat's Last Theorem, Simon Singh, 1997, ISBN 1-85702-521-0
131. Frey G (1986). "Links between stable elliptic curves and certain diophantine equations". Ann. Univ. Sarav. Ser. Math. 1: 1–40.
132. Singh, pp. 194–198; Aczel, pp. 109–114.
133. Ribet, Ken (1990). "On modular representations of Gal(Q/Q) arising from modular forms" (PDF). Inventiones mathematicae. 100 (2): 431–476. doi:10.1007/BF01231195. MR 1047143.
134. Singh, p. 205; Aczel, pp. 117–118.
135. Singh, pp. 237–238; Aczel, pp. 121–122.
136. Singh, pp. 239–243; Aczel, pp. 122–125.
137. Singh, pp. 244–253; Aczel, pp. 1–4, 126–128.
138. Aczel, pp. 128–130.
139. Singh, p. 257.
140. Singh, pp. 269–274.
141. Singh, pp. 275–277; Aczel, pp. 132–134.
142. Wiles, Andrew (1995). "Modular elliptic curves and Fermat's Last Theorem" (PDF). Annals of Mathematics. 141 (3): 443–551. doi:10.2307/2118559. JSTOR 2118559. OCLC 37032255.
143. Taylor R, Wiles A (1995). "Ring theoretic properties of certain Hecke algebras". Annals of Mathematics. 141 (3): 553–572. doi:10.2307/2118560. JSTOR 2118560. OCLC 37032255. Archived from the original on 27 November 2001.
144. Diamond, Fred (1996). "On deformation rings and Hecke rings". Annals of Mathematics. Second Series. 144 (1): 137–166. doi:10.2307/2118586. ISSN 0003-486X. MR 1405946.
145. Conrad, Brian; Diamond, Fred; Taylor, Richard (1999). "Modularity of certain potentially Barsotti-Tate Galois representations". Journal of the American Mathematical Society. 12 (2): 521–567. doi:10.1090/S0894-0347-99-00287-8. ISSN 0894-0347. MR 1639612.
146. Breuil, Christophe; Conrad, Brian; Diamond, Fred; Taylor, Richard (2001). "On the modularity of elliptic curves over Q: wild 3-adic exercises". Journal of the American Mathematical Society. 14 (4): 843–939. doi:10.1090/S0894-0347-01-00370-8. ISSN 0894-0347. MR 1839918.

147. Lenstra, Jr. H.W. (1992). On the inverse Fermat equation, Discrete Mathematics, 106–107, pp. 329–331.

148. Newton, M., "A radical diophantine equation", Journal of Number Theory 13 (1981), 495–498.

149. Bennett, Curtis D.; Glass, A. M. W.; Székely, Gábor J. (2004). "Fermat's last theorem for rational exponents". American Mathematical Monthly. 111 (4): 322–329. doi:10.2307/4145241. MR 2057186.

150. Dickson, pp. 688–691

151. Voles, Roger (July 1999). "Integer solutions of a−2 + b−2 = d−2". Mathematical Gazette. 83: 269–271.

152. Richinick, Jennifer (July 2008). "The upside-down Pythagorean Theorem". Mathematical Gazette. 92: 313–317.

153. Aczel, p. 69; Singh, p. 105.

154. Aczel, p. 69.

155. Koshy T (2001). Elementary number theory with applications. New York: Academic Press. p. 544. ISBN 978-0-12-421171-1.

156. Singh, pp. 120–125, 131–133, 295–296; Aczel, p. 70.

157. Singh, pp. 120–125.

158. Singh, p. 284

159. "The Abel Prize citation 2016" (PDF). The Abel Prize. The Abel Prize Committee. March 2016. Retrieved 16 March 2016.

160. Singh, p. 295.

161. Singh, pp. 295–296.

162. "Poincaré, Jules Henri". The American Heritage Dictionary of the English Language (fourth ed.). Boston: Houghton Mifflin Company. 2000. ISBN 0-395-82517-2. Retrieved 2007-05-05..

163. Hamilton, Richard S. (1997). "Four-manifolds with positive isotropic curvature". Communications in Analysis and Geometry. 5 (1): 1–92. MR 1456308. Zbl 0892.53018.

164. "Prize for Resolution of the Poincaré Conjecture Awarded to Dr. Grigoriy Perelman" (PDF) (Press release). Clay Mathematics Institute. March 18, 2010. Retrieved November 13, 2015. The Clay Mathematics Institute (CMI) announces today that Dr. Grigoriy Perelman of St. Petersburg, Russia, is the recipient of the Millennium Prize for resolution of the Poincaré conjecture.

165. "Последнее "нет" доктора Перельмана" The last "no" Dr. Perelman. Interfax (in Russian). July 1, 2010. Retrieved 5 April 2016.Google Translated archived link at (archived 2014-04-20)

166. Ritter, Malcolm (1 July 2010). "Russian mathematician rejects million prize". The Boston Globe.

167. Mackenzie, Dana (2006-12-22). "The Poincaré Conjecture--Proved". Science.

American Association for the Advancement of Science. 314 (5807): 1848–1849. doi:10.1126/science.314.5807.1848. ISSN 0036-8075. PMID 17185565. Archived from the original on 2007-01-02.

168. Bing, RH (1958). "Necessary and sufficient conditions that a 3-manifold be S3". Annals of Mathematics. Second Series. 68 (1): 17–37. doi:10.2307/1970041. JSTOR 1970041.
169. Bing, RH (1964). "Some aspects of the topology of 3-manifolds related to the Poincaré conjecture". Lectures on Modern Mathematics. II. New York: Wiley. pp. 93–128.
170. Milnor, John (2004). "The Poincaré Conjecture 99 Years Later: A Progress Report" (PDF). Retrieved 2007-05-05.
171. Taubes, Gary (July 1987). "What happens when hubris meets nemesis". Discover. 8: 66–77.
172. Matthews, Robert (9 April 2002). "$1 million mathematical mystery "solved"". NewScientist.com. Retrieved 2007-05-05.
173. Szpiro, George (July 29, 2008). Poincaré's Prize: The Hundred-Year Quest to Solve One of Math's Greatest Puzzles. Plume. ISBN 978-0-452-28964-2.
174. Morgan, John W., Recent progress on the Poincaré conjecture and the classification of 3-manifolds. Bull. Amer. Math. Soc. (N.S.) 42 (2005), no. 1, 57–78
175. Hamilton, Richard (1982). "Three-manifolds with positive Ricci curvature". Journal of Differential Geometry. 17 (2): 255–306. MR 0664497. Zbl 0504.53034. Reprinted in: Cao, H. D.; Chow, B.; Chu, S. C.; Yau, S.-T., eds. (2003). Collected Papers on Ricci Flow. Series in Geometry and Topology. 37. Somerville, MA: International Press. pp. 119–162. ISBN 1-57146-110-8.
176. Perelman, Grigori (2002). "The entropy formula for the Ricci flow and its geometric applications". arXiv:math.DG/0211159 .
177. Perelman, Grigori (2003). "Ricci flow with surgery on three-manifolds". arXiv:math.DG/0303109 .
178. Perelman, Grigori (2003). "Finite extinction time for the solutions to the Ricci flow on certain three-manifolds". arXiv:math.DG/0307245 .
179. Kleiner, Bruce; John W. Lott (2006). "Notes on Perelman's Papers". Geometry and Topology. 12 (5): 2587–2855. arXiv:math.DG/0605667 . doi:10.2140/gt.2008.12.2587.
180. Cao, Huai-Dong; Xi-Ping Zhu (June 2006). "A Complete Proof of the Poincaré and Geometrization Conjectures – application of the Hamilton-Perelman theory of the Ricci flow" (PDF). Asian Journal of Mathematics. 10 (2).
181. Cao, Huai-Dong & Zhu, Xi-Ping (December 3, 2006). "Hamilton–Perelman's Proof of the Poincaré Conjecture and the Geometrization Conjecture". arXiv:math.DG/0612069 .
182. Morgan, John; Gang Tian (2006). "Ricci Flow and the Poincaré Conjecture". arXiv:math.DG/0607607 .
183. Morgan, John; Gang Tian (2007). Ricci Flow and the Poincaré Conjecture. Clay

Mathematics Institute. ISBN 0-8218-4328-1.

184. Nasar, Sylvia; David Gruber (August 28, 2006). "Manifold destiny". The New Yorker. pp. 44–57. On-line version at the New Yorkerwebsite.
185. Chang, Kenneth (August 22, 2006). "Highest Honor in Mathematics Is Refused". New York Times.
186. "Reclusive Russian solves 100-year-old maths problem". China Daily. 23 August 2006. p. 7.
187. A Report on the Poincaré Conjecture. Special lecture by John Morgan.
188. Terence Tao wrote an exposition of Ricci flow with surgery in: Tao, Terence (2006). "Perelman's proof of the Poincaré conjecture: a nonlinear PDE perspective". arXiv:math.DG/0610903 .
189. Prize for Resolution of the Poincaré Conjecture Awarded to Dr. Grigoriy Perelman
190. Malcolm Ritter (2010-07-01). "Russian mathematician rejects $1 million prize". AP on PhysOrg. Retrieved 2011-05-15.
191. http://arxiv.org/pdf/1303.2673.pdf DIVISION-ALGEBRAS/POINCARE-COJECTURE CORRESPONDENCE
192. McKee, Maggie (14 May 2013). "First proof that infinitely many prime numbers come in pairs". Nature. doi:10.1038/nature.2013.12989. ISSN 0028-0836.
193. McKee, M. (2013). "First proof that prime numbers pair up into infinity". Nature. doi:10.1038/nature.2013.12989.
194. Zhang, Yitang (2014). "Bounded gaps between primes". Annals of Mathematics. 179 (3): 1121–1174. doi:10.4007/annals.2014.179.3.7. MR 3171761.
195. Tao, Terence (June 4, 2013). "Polymath proposal: bounded gaps between primes".
196. "Bounded gaps between primes". Polymath. Retrieved 2014-03-27.
197. "Bounded gaps between primes". Polymath. Retrieved 2014-02-21.
198. https://primes.utm.edu/lists/small/100ktwins.txt
199. Brun, V. (1915), "Über das Goldbachsche Gesetz und die Anzahl der Primzahlpaare", Archiv for Mathematik og Naturvidenskab (in German), 34 (8): 3–19, ISSN 0365-4524, JFM 45.0330.16
200. Bateman & Diamond (2004) p. 313
201. Heini Halberstam, and Hans-Egon Richert, Sieve Methods, p. 117, Dover Publications, 2010
202. Goldston, Daniel Alan; Motohashi, Yoichi; Pintz, János; Yıldırım, Cem Yalçın (2006), "Small gaps between primes exist", Japan Academy. Proceedings. Series A. Mathematical Sciences, 82 (4): 61–65, arXiv:math.NT/0505300 , doi:10.3792/pjaa.82.61, MR 2222213.
203. Goldston, D. A.; Graham, S. W.; Pintz, J.; Yıldırım, C. Y. (2009), "Small gaps between primes or almost primes", Transactions of the American Mathematical Society, 361 (10): 5285–5330, arXiv:math.NT/0506067 , doi:10.1090/S0002-9947-09-04788-6, MR 2515812.

204. "A page of number theoretical constants". 2007. Retrieved 2011-02-02.

205. Bateman & Diamond (2004) pp.334–335

206. Caldwell, Chris K. "The Prime Database: 2996863034895*2^1290000-1".

207. "World Record Twin Primes Found!".

208. Tomás Oliveira e Silva (7 April 2008). "Tables of values of pi(x) and of pi2(x)". Aveiro University. Retrieved 7 January 2011.

209. Maddux, Cleborne D.; Johnson, D. Lamont (1997). Logo: A Retrospective. New York: Haworth Press. p. 160. ISBN 0-7890-0374-0. The problem is also known by several other names, including: Ulam's conjecture, the Hailstone problem, the Syracuse problem, Kakutani's problem, Hasse's algorithm, and the Collatz problem.

210. Lagarias, Jeffrey C. (1985). "The 3x + 1 problem and its generalizations". The American Mathematical Monthly. 92 (1): 3–23. JSTOR 2322189.

211. According to Lagarias (1985), p. 4, the name "Syracuse problem" was proposed by Hasse in the 1950s, during a visit to Syracuse University.

212. Pickover, Clifford A. (2001). Wonders of Numbers. Oxford: Oxford University Press. pp. 116–118. ISBN 0-19-513342-0.

213. "Hailstone Number". MathWorld. Wolfram Research.

214. Hofstadter, Douglas R. (1979). Gödel, Escher, Bach. New York: Basic Books. pp. 400–2. ISBN 0-465-02685-0.

215. Friendly, Michael (1988). Advanced Logo: A Language for Learning. Hillsdale NJ: Lawrence Erlbaum Associates. ISBN 0-89859-933-4.

216. Guy, Richard K. (2004). ""E17: Permutation Sequences"". Unsolved problems in number theory (3rd ed.). Springer-Verlag. pp. 336–7. ISBN 0-387-20860-7. Zbl 1058.11001.

217. Guy, R. K. (1983). "Don't try to solve these problems". Amer. Math. Monthly. 90: 35–41. JSTOR 2975688. By this Erdos means that there aren't powerful tools for manipulating such objects.

218. Lagarias, Jeffrey C., ed. (2010). The ultimate challenge: the 3x+1 problem. Providence, R.I.: American Mathematical Society. p. 4. ISBN 0821849409.

219. Leavens, Gary T.; Vermeulen, Mike (December 1992). "3x+1 Search Programs". Computers & Mathematics with Applications. 24 (11): 79–99. doi:10.1016/0898-1221(92)90034-F.

220. Roosendaal, Eric. "3x+1 Delay Records". Retrieved 27 November 2011. (Note: "Delay records" are total stopping time records.)

221. Simons, J.; de Weger, B. (2003). "Theoretical and computational bounds for m-cycles of the 3n + 1 problem" (PDF). Acta Arithmetica. 117 (1): 51–70. doi:10.4064/aa117-1-3.

222. Steiner, R. P. (1977). "A theorem on the syracuse problem". Proceedings of the 7th Manitoba Conference on Numerical Mathematics. pp. 553–9.

223. Simons, John L. (2005). "On the nonexistence of 2-cycles for the 3x+1 problem".

Math. Comp. 74: 1565–72. doi:10.1090/s0025-5718-04-01728-4. MR 2137019.

224. Silva, Tomás Oliveira e Silva. "Computational verification of the 3x+1 conjecture". Retrieved 13 May 2015.

225. Garner, Lynn E. "On The Collatz 3n + 1 Algorithm" (PDF). Retrieved 27 March 2015.

226. Lagarias (1985), section "A heuristic argument".

227. Krasikov, Ilia; Lagarias, Jeffrey C. (2003). "Bounds for the 3x + 1 problem using difference inequalities". Acta Arithmetica. 109 (3): 237–258. doi:10.4064/aa109-3-4. MR 1980260.

228. Colussi, Livio (9 September 2011). "The convergence classes of Collatz function". Theoretical Computer Science. 412 (39): 5409–5419. doi:10.1016/j.tcs.2011.05.056.

229. Hew, Patrick Chisan (7 March 2016). "Working in binary protects the repetends of 1/3h: Comment on Colussi's 'The convergence classes of Collatz function'". Theoretical Computer Science. 618: 135–141. doi:10.1016/j.tcs.2015.12.033.

230. Terras, Riho (1976), "A stopping time problem on the positive integers" (PDF), Polska Akademia Nauk, 30 (3): 241–252, MR 0568274

231. Chamberland, Marc (1996). "A continuous extension of the 3x + 1 problem to the real line". Dynam. Contin. Discrete Impuls Systems. 2 (4): 495–509.

232. Letherman, Simon; Schleicher, Dierk; Wood, Reg (1999). "The (3n + 1)-Problem and Holomorphic Dynamics". Experimental Mathematics. 8 (3): 241–252. doi:10.1080/10586458.1999.10504402.

233. Scollo, Giuseppe (2007), "Looking for Class Records in the 3x+1 Problem by means of the COMETA Grid Infrastructure", Grid Open Days at the University of Palermo (PDF)

234. Lagarias (1985), Theorem D.

235. Conway, John H. (1972), "Unpredictable Iterations", Proc. 1972 Number Theory Conf., Univ. Colorado, Boulder, pp. 49–52

236. Kurtz, Stuart A.; Simon, Janos (2007). "The Undecidability of the Generalized Collatz Problem". In Cai, J.-Y.; Cooper, S. B.; Zhu, H. Proceedings of the 4th International Conference on Theory and Applications of Models of Computation, TAMC 2007, held in Shanghai, China in May 2007. pp. 542–553. doi:10.1007/978-3-540-72504-6_49. ISBN 3-540-72503-2. As PDF

237. Ben-Amram, Amir M. (2015), "Mortality of iterated piecewise affine functions over the integers: Decidability and complexity", Computability, 1 (1): 19–56, doi:10.3233/COM-150032

238. Lagarias, Jeffrey C., ed. (2010). The Ultimate Challenge: the 3x+1 problem. American Mathematical Society. ISBN 978-0-8218-4940-8. Zbl 1253.11003.

239. Milnor, John W. (1961). "Two complexes which are homeomorphic but combinatorially distinct". Annals of Mathematics. 74 (2): 575–590. doi:10.2307/1970299. JSTOR 1970299. MR 133127.

240. Moise, Edwin E. (1977). Geometric Topology in Dimensions 2 and 3. New York: New York : Springer-Verlag. ISBN 978-0-387-90220-3.
241. Hamilton, Richard S. (1997). "Four-manifolds with positive isotropic curvature". Communications in Analysis and Geometry. 5 (1): 1–92. MR 1456308. Zbl 0892.53018.
242. Juris Hartmanis 1989, Gödel, von Neumann, and the P = NP problem, Bulletin of the European Association for Theoretical Computer Science, vol. 38, pp. 101–107
243. Cook, Stephen (1971). "The complexity of theorem proving procedures". Proceedings of the Third Annual ACM Symposium on Theory of Computing. pp. 151–158.
244. Lance Fortnow, The status of the P versus NP problem, Communications of the ACM 52 (2009), no. 9, pp. 78–86. doi:10.1145/1562164.1562186
245. Langlands, Robert (1967), Letter to Prof. Weil
246. Popper, Karl (2004). Conjectures and refutations : the growth of scientific knowledge. London: Routledge. ISBN 0-415-28594-1
247. de Haro et al. 2013, p. 2
248. Klebanov and Maldacena 2009
249. "Top Cited Articles of All Time (2014 edition)". INSPIRE-HEP. Retrieved 26 December 2015.
250. A standard textbook on general relativity is Wald 1984.
251. Maldacena 2005, p. 58
252. Griffiths 2004
253. Maldacena 2005, p. 62
254. See the subsection entitled "Examples of the correspondence". For examples which do not involve string theory or M-theory, see the section entitled "Generalizations".
255. Wald 1984, p. 4
256. Zwiebach 2009, p. 8
257. Zwiebach 2009, pp. 7–8
258. This analogy is used for example in Greene 2000, p. 186.
259. A standard text is Peskin and Schroeder 1995.
260. For an introduction to the applications of quantum field theory to condensed matter physics, see Zee 2010.
261. Conformal field theories are characterized by their invariance under conformal transformations.
262. For an introduction to conformal field theory emphasizing its applications to perturbative string theory, see Volume II of Deligne et al. 1999.
263. Klebanov and Maldacena 2009, p. 28
264. Maldacena 2005, p. 60

265. Maldacena 2005, p. 61
266. The mathematical relationship between the interior and boundary of anti-de Sitter space is related to the ambient construction of Charles Fefferman and Robin Graham. For details see Fefferman and Graham 1985, Fefferman and Graham 2011.
267. Zwiebach 2009, p. 552
268. Maldacena 2005, pp. 61–62
269. Maldacena 2005, p. 57
270. The known realizations of AdS/CFT typically involve unphysical numbers of spacetime dimensions and unphysical supersymmetries.
271. This example is the main subject of the three pioneering articles on AdS/CFT: Maldacena 1998; Gubser, Klebanov, and Polyakov 1998; and Witten 1998.
272. Merali 2011, p. 303; Kovtun, Son, and Starinets 2001
273. Maldacena 1998
274. For a review of the (2,0)-theory, see Moore 2012.
275. See Moore 2012 and Alday, Gaiotto, and Tachikawa 2010.
276. Aharony et al. 2008
277. Aharony et al. 2008, sec. 1
278. A standard textbook introducing the formalism of Feynman diagrams is Peskin and Schroeder 1995.
279. Zee 2010, p. 43
280. Zwiebach 2009, p. 12
281. Maldacena 1998, sec. 6
282. Hawking 1975
283. For an accessible introduction to the black hole information paradox, and the related scientific dispute between Hawking and Leonard Susskind, see Susskind 2008.
284. Zwiebach 2009, p. 554
285. Maldacena 2005, p. 63
286. Hawking 2005
287. Zwiebach 2009, p. 559
288. More precisely, one cannot apply the methods of perturbative quantum field theory.
289. Zwiebach 2009, p. 561; Kovtun, Son, and Starinets 2001
290. Merali 2011, p. 303; Luzum and Romatschke 2008
291. Zwiebach 2009, p. 561
292. Merali 2011
293. Merali 2011, p. 303

294. Sachdev 2013, p. 51
295. McLerran 2007
296. Anderson, Philip. "Strange connections to strange metals". Physics Today. Retrieved 14 August 2013.
297. Zwiebach 2009, p. 525
298. Aharony et al. 2008, sec. 1.1
299. Scherk and Schwarz 1974
300. 't Hooft 1974
301. Bekenstein 1973
302. Susskind 2008
303. 't Hooft 1993
304. Susskind 1995
305. Susskind 2008, p. 444
306. Polyakov 2008, p. 6
307. Gubser, Klebanov, and Polyakov 1998
308. Witten 1998
309. Maldacena 2005, p. 63; Cowen 2013
310. Merali 2011, pp. 302–303
311. Merali 2011; Sachdev 2013
312. Rangamani 2009
313. For a review, see Carlip 2003.
314. According to the results of Witten 1988, three-dimensional quantum gravity can be understood by relating it to Chern–Simons theory.
315. Brown and Henneaux 1986
316. Coussaert, Henneaux, and van Driel 1995
317. Witten 2007
318. Guica et al. 2009, p. 1
319. Perlmutter 2003
320. Biquard 2005, p. 33
321. Strominger 2001
322. See the subsection entitled "Black hole information paradox".
323. Guica et al. 2009
324. Castro, Maloney, and Strominger 2010
325. Klebanov and Polyakov 2002
326. See the Introduction in Klebanov and Polyakov 2002.
327. Giombi and Yin 2010

CHAPTER 6

Misnamed Theorems

Chapter Contents

INTRODUCTION

This is a list of misnamed theorems in mathematics. It includes theorems (and lemmas, corollaries, conjectures, laws, and perhaps even the odd object) that are well known in mathematics, but which are not named for the originator. That is, these items on this list illustrate Stigler's law of eponymy (which is not, of course, due to Stephen Stigler, who credits Robert K Merton).

- Benford's law. This was first stated in 1881 by Simon Newcomb, and rediscovered in 1938 by Frank Benford. The first rigorous formulation and proof seems to be due to Ted Hill in 1988.
- Bertrand's ballot theorem. This result concerning the probability that the winner of an election was ahead at each step of ballot counting was first published by

W. A. Whitworth in 1878, but named after Joseph Louis François Bertrand who rediscovered it in 1887.

- Bézout's theorem. The statement may have been made first by Isaac Newton in 1665. The matter of a proof was taken up by Colin MacLaurin (c. 1720) and Leonhard Euler as well as Étienne Bézout (c. 1750). However, Bézout's "proof" was *incorrect*. The first correct proof seems to be due mostly to Georges-Henri Halphen in the 1870s.
- Burnside's lemma. This was stated and proved without attribution in Burnside's 1897 textbook, but it had previously been discussed by Augustin Cauchy, in 1845, and by Georg Frobenius in 1887.
- Cayley–Hamilton theorem. The theorem was first proved in the easy special case of 2×2 matrices by Cayley, and later for the case of 4×4 matrices by Hamilton. But it was only proved in general by Frobenius in 1878.
- Cramer's paradox. This was first noted by Colin Maclaurin in 1720, and then rediscovered by Leonhard Euler in 1748 (whose paper was not published for another two years, as Euler wrote his papers faster than his printers could print them). It was also discussed by Gabriel Cramer in 1750, who independently suggested the essential idea needed for the resolution, although providing a rigorous proof remained an outstanding open problem for much of the 19th century. Even though Cramer had cited Maclaurin, the paradox became known after Cramer rather than Maclaurin. Halphen, Arthur Cayley, and several other luminaries contributed to the earliest more or less correct proof. See for an excellent review.
- Cramer's rule. It is named after Gabriel Cramer (1704–1752), who published the rule in his 1750 *Introduction à l'analyse des lignes courbes algébriques*, although Colin Maclaurin also published the method in his 1748 *Treatise of Algebra* (and probably knew of the method as early as 1729).
- Frobenius theorem. This fundamental theorem was stated and proved in 1840 by Feodor Deahna. Even though Frobenius cited Deahna's paper in his own 1875 paper, it became known after Frobenius, not Deahna. See for a historical review.
- Heine–Borel theorem. This theorem was proved in 1872 by Émile Borel, not by Eduard Heine. Borel used techniques similar to those that Heine used to prove that continuous functions on closed intervals are uniformly continuous. Heine's name was attached because Schönflies noticed the similarity in Heine's and Borel's approaches. In fact, the theorem was first proved in 1852 by Peter Gustav Lejeune Dirichlet, but Lejeune Dirichlet's lecture notes were not published until 1904.
- L'Hôpital's rule. This rule first appeared in l'Hôpital's book *L'Analyse des Infiniment Petits pour l'Intelligence des Lignes Courbes* in 1696. The rule is believed to be the work of Johann Bernoulli since l'Hôpital, a nobleman, paid Bernoulli a retainer of 300 francs per year to keep him updated on developments in calculus and to solve problems he had. See L'Analyse des Infiniment Petits pour l'Intelligence des Lignes Courbes and reference therein.

- Maclaurin series. The Maclaurin series was named after Colin Maclaurin, a professor in Edinburgh, who published this special case of the Taylor series in 1742, but never claimed to have discovered it.
- Marden's theorem. This theorem relating the location of the zeros of a complex cubic polynomial to the zeros of its derivative was named by Dan Kalman after Kalman read it in a 1966 book by Morris Marden, who had first written about it in 1945. But, as Marden had himself written, its original proof was by Jörg Siebeck in 1864.
- Morrie's law. The name is due to physicist Richard Feynman, who used to refer to the identity under that name. Feynman picked that name because he had learned the law during his childhood from a boy with the name Morrie Jacobs.
- Pell's equation. The solution of the equation $x^2 - dy^2 = 1$, where x and y are unknown positive integers and where d is a known positive integer which is not a perfect square, is ascribed to John Pell. It seems to have been discovered by Fermat, who set it as a challenge problem in 1657. The first European solution is found in a joint work in 1658 by John Wallis and Lord Brouncker; in 1668, a shorter solution was given in an edition of a third mathematician's work by Pell; see ref. The first rigorous proof may be due to Lagrange. The misnomer apparently came about when Euler confused Brouncker and Pell; see for an extensive account of the history of this equation.
- Poincaré lemma. This was mentioned in 1886 by Henri Poincaré, but was first proved in a series of 1889 papers by the distinguished Italian mathematician Vito Volterra. Nonetheless it has become known after Poincaré. See for the twisted history of this lemma.
- Pólya enumeration theorem. This was proven in 1927 in a difficult paper by J. H. Redfield. Despite the prominence of the venue (the American Journal of Mathematics), the paper was overlooked. Eventually, the theorem was independently rediscovered in 1936 by George Pólya. Not until 1960 did Frank Harary unearth the much earlier paper by Redfield. See for historical and other information.
- Stokes' theorem. It is named after Sir George Gabriel Stokes (1819–1903), although the first known statement of the theorem is by William Thomson (Lord Kelvin) and appears in a letter of his to Stokes. The theorem acquired its name from Stokes' habit of including it in the Cambridge prize examinations. In 1854 he asked his students to prove the theorem in an examination; it is not known if anyone was able to do so.
- Zorn's lemma is named for Max Zorn. Much work on the theorem now known as Zorn's lemma, and on several closely related formulations such as the Hausdorff maximal principle, was done between 1907 and 1940 by Zorn, Brouwer, Hausdorff, Kuratowski, R. L. Moore, and others. But the particular theorem now known as "Zorn's lemma" was never proved by Zorn, and in any event Zorn's results were anticipated by Kuratowski. The theorem was discovered by Chevalley in 1936, and published and attributed to Zorn by him in Bourbaki's

Théorie des Ensembles in 1939. A very similar result was anticipated by S. Bochner in 1928.

REFERENCES

1. Newcomb, S. (1881). "Note on the frequency of use of the different digits in natural numbers". Amer. J. Math. The Johns Hopkins University Press. 4 (1): 39–40. doi:10.2307/2369148. JSTOR 2369148.
2. Benford, F. (1938). "The law of anomalous numbers". Proc. Amer. Phil. Soc. 78: 551–572.
3. Hill, Theodore P. (April 1995). "The Significant Digit Phenomenon". Am. Math. Monthly. Mathematical Association of America. 102 (4): 322–327. doi:10.2307/2974952. JSTOR 2974952.
4. Feller, William (1968), An Introduction to Probability Theory and its Applications, Volume I (3rd ed.), Wiley, p. 69.
5. Bix, Robert (1998). Conics and Cubics. Springer. ISBN 0-387-98401-1.
6. Burnside, William (1897). Theory of groups of finite order. Cambridge University Press.
7. Grattan-Guinness, Ivor (2002), Companion Encyclopaedia of the History and Philosophy of the Mathematical Sciences, Routledge, pp. 779–780, ISBN 9781134957507.
8. Scott, Charlotte Agnas (March 1898). "On the Intersection of Plane Curves". Bull. Am. Math. Soc. 4 (6): 260–273. doi:10.1090/S0002-9904-1898-00489-5.
9. Carl B. Boyer (1968). A History of Mathematics, 2nd edition. Wiley. p. 431.
10. Deahna, F. (1840). "Über die Bedingungen der Integrabilität". J. Reine Angew. Math. 20: 340–350.
11. Frobenius, Georg (1895). "Ūber das Pfaffsche Problem". J. Reine Angew. Math.: 230–315.
12. Samelson, Hans (June–July 2001). "Differential Forms, the Early days; or the Stories of Deahna's Theorem and of Volterra's Theorem". Am. Math. Monthly. Mathematical Association of America. 108 (6): 552–530. doi:10.2307/2695706. JSTOR 2695706.
13. Sundström, Manya Raman (2010). "A pedagogical history of compactness". arXiv:1006.4131v1.
14. Thomas & Finney. Calculus and Analytic Geometry.
15. Kalman, Dan (2008), "An Elementary Proof of Marden's Theorem", The American Mathematical Monthly, 115: 330–338, ISSN 0002-9890
16. Siebeck, Jörg (1864), "Über eine neue analytische Behandlungweise der Brennpunkte", Journal für die reine und angewandte Mathematik, 64: 175–182, ISSN 0075-4102

17. W.A. Beyer, J.D. Louck, and D. Zeilberger, A Generalization of a Curiosity that Feynman Remembered All His Life, Math. Mag. 69, 43–44, 1996.
18. Cajori, Florian (1999). A History of Mathematics. New York: Chelsea. ISBN 0-8284-0203-5. (reprint of fifth edition, 1891).
19. Whitford, Edward Everett (1912). The Pell Equation. New York: E. E. Whitford. This is Whitford's 1912 Ph.D. dissertation, written at Columbia University and published at his own expense in 1912.
20. Poincaré, H. (1886–1887). "Sur les residus des intégrales doubles". Acta Math. 9: 321–380. doi:10.1007/BF02406742.
21. Redfield, J. H. (1927). "The theory of group related distributions". Amer. J. Math. The Johns Hopkins University Press. 49 (3): 433–445. doi:10.2307/2370675. JSTOR 2370675.
22. Pólya, G. (1936). "Algebraische Berechnung der Isomeren einiger organischer Verbindungen". Zeitschrift für Kristallographie A. 93: 414. doi:10.1524/zkri.1936.93.1.415.
23. Read, R. C. (December 1987). "Pólya's Theorem and its Progeny". Mathematics Magazine. 60 (5): 275–282. doi:10.2307/2690407. JSTOR 2690407.
24. Victor J. Katz (May 1979). "The History of Stokes' Theorem". Mathematics Magazine. 52 (3): 146–156. doi:10.2307/2690275. JSTOR 2690275.
25. Campbell, Paul J. (1978). "The Origin of ‹Zorn's Lemma›". Historia Mathematica. 5: 77–89. doi:10.1016/0315-0860(78)90136-2.

CHAPTER 7

Proposition

Chapter Contents

- **Introduction**
- **By Aristotle**
- **Objections to Propositions**

A proposition is a proposed plan of action, a detailed suggestion. You might consider your friend's suggestion to set up a neighborhood snow-shoveling business after the blizzard a winning proposition. The noun proposition means something presented for consideration. In some contexts, that something being offered is sexual. Therefore, don't use the term proposition loosely. If you say that you made a proposition to your teacher, people could misinterpret you. It's safer to stick with suggestion.

INTRODUCTION

The term proposition has a broad use in contemporary philosophy. It is used to refer to some or all of the following: the primary bearers of truth-value, the objects of belief and other "propositional attitudes" (i.e., what is believed, doubted, etc.), the referents of that-clauses, and the meanings of declarative sentences. Propositions are the sharable objects of attitudes and the primary bearers of truth and falsity. This stipulation rules out certain candidates for propositions, including thought- and utterance-tokens which are not sharable, and concrete events or facts, which cannot be false.

BY ARISTOTLE

Aristotelian logic identifies a proposition as a sentence which affirms or denies a predicate of a subject. An Aristotelian proposition may take the form "All men are mortal" or "Socrates is a man." In the first example the subject is "All men" and the predicate "are mortal." In the second example the subject is "Socrates" and the predicate is "is a man."

By the Logical Positivists

Often propositions are related to closed sentences to distinguish them from what is expressed by an open sentence. In this sense, propositions are "statements" that are truth-bearers. This conception of a proposition was supported by the philosophical school of logical positivism.

Some philosophers argue that some (or all) kinds of speech or actions besides the declarative ones also have propositional content. For example, yes–no questions present propositions, being inquiries into the truth value of them. On the other hand, some signs can be declarative assertions of propositions without forming a sentence nor even being linguistic, e.g. traffic signs convey definite meaning which is either true or false.

Propositions are also spoken of as the content of beliefs and similar intentional attitudes such as desires, preferences, and hopes. For example, "I desire *that I have a new car*," or "I wonder *whether it will snow*" (or, whether it is the case that "it will snow"). Desire, belief, and so on, are thus called propositional attitudes when they take this sort of content.

By Russell

Bertrand Russell held that propositions were structured entities with objects and properties as constituents. Wittgenstein held that a proposition is the set of possible worlds/states of affairs in which it is true. One important difference between these views is that on the Russellian account, two propositions that are true in all the same states of affairs can still be differentiated. For instance, the proposition that two plus two equals four is distinct on a Russellian account from three plus three equals six. If propositions are sets of possible worlds, however, then all mathematical truths (and all other necessary truths) are the same set (the set of all possible worlds).

Relation to the Mind

In relation to the mind, propositions are discussed primarily as they fit into propositional attitudes. Propositional attitudes are simply attitudes characteristic of folk psychology (belief, desire, etc.) that one can take toward a proposition (e.g. 'it is raining,' 'snow is white,' etc.). In English, propositions usually follow folk psychological attitudes by a "that clause" (e.g. "Jane believes *that* it is raining"). In philosophy of mind and psychology, mental states are often taken to primarily consist in propositional attitudes. The propositions are usually said to be the "mental content" of the attitude. For example, if Jane has a mental state of believing that it is raining, her mental content is the proposition 'it is raining.' Furthermore, since such mental states are *about* something (namely propositions), they are said to be intentional mental states. Philosophical debates surrounding propositions as they relate to propositional attitudes have also recently centered on whether they are internal or external to the agent or whether they are mind-dependent or mind-independent entities (see the entry on internalism and externalism in philosophy of mind).

Treatment in Logic

As noted above, in Aristotelian logic a proposition is a particular kind of sentence, one which affirms or denies a predicate of a subject. Aristotelian propositions take forms like "All men are mortal" and "Socrates is a man."

Propositions show up in modern formal logic as objects of a formal language. A formal language begins with different types of symbols. These types can include variables, operators, function symbols, predicate (or relation) symbols, quantifiers, and propositional constants. (Grouping symbols are often added for convenience in using the language but do not play a logical role.) Symbols are concatenated together according to recursive rules in order to construct strings to which truth-values will be assigned. The rules specify how the operators, function and predicate symbols, and quantifiers are to be concatenated with other strings. A proposition is then a string with a specific form. The form that a proposition takes depends on the type of logic.

The type of logic called propositional, sentential, or statement logic includes only operators and propositional constants as symbols in its language. The propositions in this language are propositional constants, which are considered atomic propositions, and composite propositions, which are composed by recursively applying operators to propositions. *Application* here is simply a short way of saying that the corresponding concatenation rule has been applied.

The types of logics called predicate, quantificational, or *n*-order logic include variables, operators, predicate and function symbols, and quantifiers as symbols in their languages. The propositions in these logics are more complex. First, terms must be defined. A term is (i) a variable or (ii) a function symbol applied to the number of terms required by the function symbol's arity. For example, if + is a binary function symbol and x, y, and z are variables, then $x+(y+z)$ is a term, which might be written with the symbols in various orders. A proposition is (i) a predicate symbol applied to the number of terms required by its arity, (ii) an operator applied to the number of propositions required by its arity, or (iii) a quantifier applied to a proposition. For example, if = is a binary predicate symbol and ∀ is a quantifier, then $\forall x,y,z\ [(x = y) \rightarrow (x+z = y+z)]$ is a proposition. This more complex structure of propositions allows these logics to make finer distinctions between inferences, i.e., to have greater expressive power.

In this context, propositions are also called sentences, statements, statement forms, formulas, and well-formed formulas, though these terms are usually not synonymous within a single text. This definition treats propositions as syntactic objects, as opposed to semantic or mental objects. That is, propositions in this sense are meaningless, formal, abstract objects. They are assigned meaning and truth-values by mappings called interpretations and valuations, respectively.

OBJECTIONS TO PROPOSITIONS

Attempts to provide a workable definition of proposition include

Two meaningful declarative sentences express the same proposition if and only if they mean the same thing.

Thus defining *proposition* in terms of synonymity. For example, "Snow is white" (in English) and "Schnee ist weiß" (in German) are different sentences, but they say the same thing, so they express the same proposition.

Two meaningful declarative sentence-tokens express the same proposition if and only if they mean the same thing.

Unfortunately, the above definitions have the result that two sentences/sentence-tokens which have the same meaning and thus express the same proposition, could have different truth-values, e.g. "I am Spartacus" said by Spartacus and said by John Smith; and e.g. "It is Wednesday" said on a Wednesday and on a Thursday.

A number of philosophers and linguists claim that all definitions of a proposition are too vague to be useful. For them, it is just a misleading concept that should be removed from philosophy and semantics. W.V. Quine maintained that the indeterminacy of translation prevented any meaningful discussion of propositions, and that they should be discarded in favor of sentences. Strawson advocated the use of the term "statement".

REFERENCES

1. "Propositions (Stanford Encyclopedia of Philosophy)". Plato.stanford.edu. Retrieved 2014-06-23.
2. Quine W.V. Philosophy of Logic, Prentice-Hall NJ USA: 1970, pp 1-14

CHAPTER 8

Inequality Mathematics

Chapter Contents

Inequalities are a set of five symbols used to demonstrate instances where one value is not the same as another value. The five symbols are described as "not equal to," "greater than," "greater than or equal to," "less than" and "less than or equal to.

INTRODUCTION

In mathematics, an **inequality** is a relation that holds between two values when they are different (see also: equality).

- The notation $a \neq b$ means that a is **not equal to** b.

It does not say that one is greater than the other, or even that they can be compared in size.

If the values in question are elements of an ordered set, such as the integers or the real numbers, they can be compared in size.

- The notation $a < b$ means that a is **less than** b.
- The notation $a > b$ means that a is **greater than** b.

In either case, a is not equal to b. These relations are known as **strict inequalities**. The notation $a < b$ may also be read as "a is strictly less than b".

In contrast to strict inequalities, there are two types of inequality relations that are not strict:

- The notation $a \leq b$ means that a is **less than or equal to** b (or, equivalently, **not greater than** b, or **at most** b); "not greater than" can also be represented by the symbol for "greater than" bisected by a vertical line, "not."
- The notation $a \geq b$ means that a is **greater than or equal to** b (or, equivalently, **not less than** b, or **at least** b),; "not less than" can also be represented by the symbol for "less than" bisected by a vertical line, "not."

In engineering sciences, a less formal use of the notation is to state that one quantity is "much greater" than another, normally by several orders of magnitude.

- The notation $a \ll b$ means that a is **much less than** b. (In measure theory, however, this notation is used for absolute continuity, an unrelated concept.)
- The notation $a \gg b$ means that a is **much greater than** b.

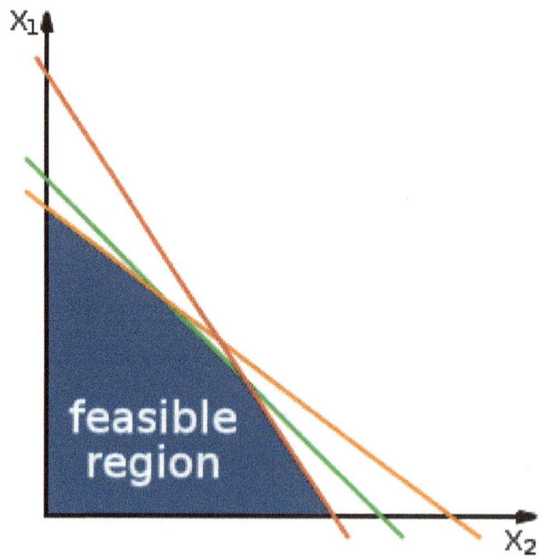

The feasible regions of linear programming are defined by a set of inequalities.

PROPERTIES

Inequalities are governed by the following properties. All of these properties also hold if all of the non-strict inequalities ($\leq$ and $\geq$) are replaced by their corresponding strict inequalities ($<$ and $>$) and (in the case of applying a function) monotonic functions are limited to *strictly* monotonic functions.

Transitivity

The transitive property of inequality states:

- For any real numbers a, b, c:
- If $a \geq b$ and $b \geq c$, then $a \geq c$.
- If $a \leq b$ and $b \leq c$, then $a \leq c$.
- If *either* of the premises is a strict inequality, then the conclusion is a strict inequality.
- E.g. if $a \geq b$ and $b > c$, then $a > c$
- An equality is of course a special case of a non-strict inequality.
- E.g. if $a = b$ and $b > c$, then $a > c$

Converse

The relations $\leq$ and $\geq$ are each other's converse:

- For any real numbers a and b:
- If $a \leq b$, then $b \geq a$.
- If $a \geq b$, then $b \leq a$.

Addition and Subtraction

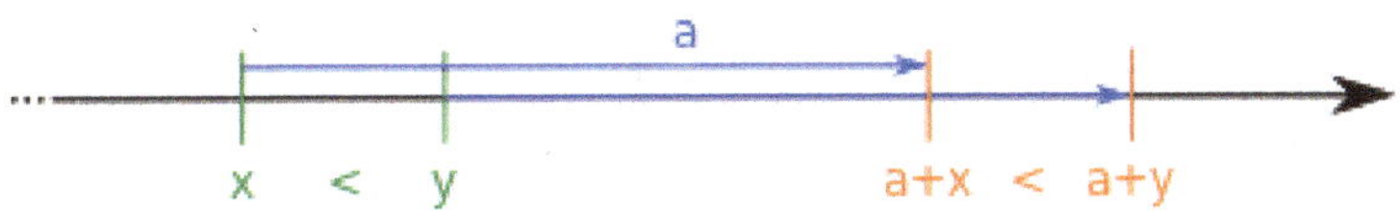

If $x < y$, then $x + a < y + a$.

A common constant c may be added to or subtracted from both sides of an inequality:

- For any real numbers a, b, c
- If $a \leq b$, then $a + c \leq b + c$ and $a - c \leq b - c$.
- If $a \geq b$, then $a + c \geq b + c$ and $a - c \geq b - c$.

i.e., the real numbers are an ordered group under addition.

Multiplication and Division

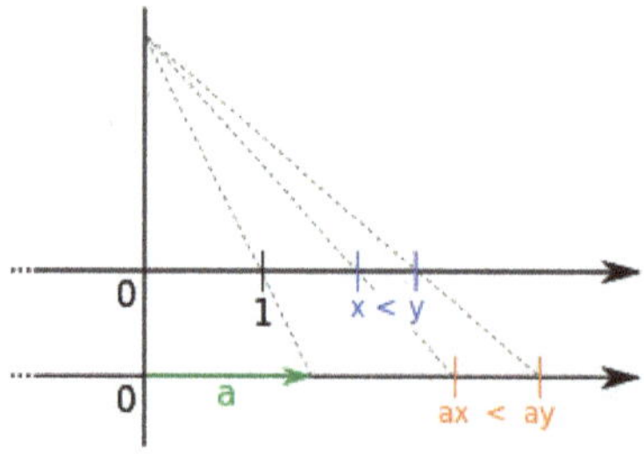

If $x < y$ and $a > 0$, then $ax < ay$.

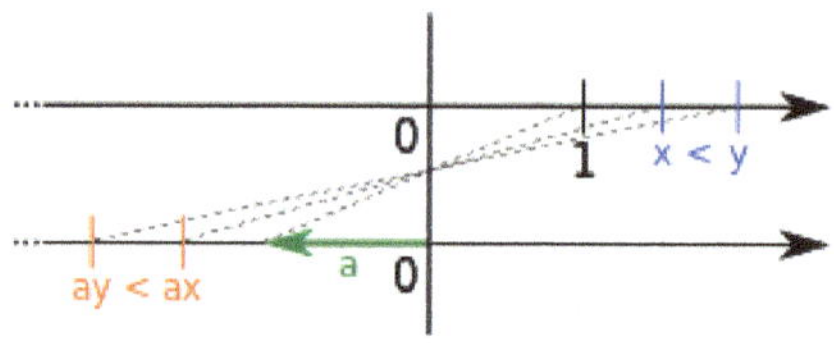

If $x < y$ and $a < 0$, then $ax > ay$.

The properties that deal with multiplication and division state:

- For any real numbers, a, b and non-zero c:
    - If c is positive, then multiplying or dividing by c does not change the inequality:
    - If $a \geq b$ and $c > 0$, then $ac \geq bc$ and $a/c \geq b/c$.
    - If $a \leq b$ and $c > 0$, then $ac \leq bc$ and $a/c \leq b/c$.
    - If c is negative, then multiplying or dividing by c inverts the inequality:
    - If $a \geq b$ and $c < 0$, then $ac \leq bc$ and $a/c \leq b/c$.
    - If $a \leq b$ and $c < 0$, then $ac \geq bc$ and $a/c \geq b/c$.

More generally, this applies for an ordered field, see below.

Additive Inverse

The properties for the additive inverse state:

- For any real numbers a and b, negation inverts the inequality:
    - If $a \leq b$, then $-a \geq -b$.
    - If $a \geq b$, then $-a \leq -b$.

Multiplicative Inverse

The properties for the multiplicative inverse state:

- For any non-zero real numbers a and b that are both positive or both negative:
  - If $a \leq b$, then $1/a \geq 1/b$.
  - If $a \geq b$, then $1/a \leq 1/b$.
- If one of a and b is positive and the other is negative, then:
  - If $a < b$, then $1/a < 1/b$.
  - If $a > b$, then $1/a > 1/b$.

These can also be written in chained notation as:

- For any non-zero real numbers a and b:
  - If $0 < a \leq b$, then $1/a \geq 1/b > 0$.
  - If $a \leq b < 0$, then $0 > 1/a \geq 1/b$.
  - If $a < 0 < b$, then $1/a < 0 < 1/b$.
  - If $0 > a \geq b$, then $1/a \leq 1/b < 0$.
  - If $a \geq b > 0$, then $0 < 1/a \leq 1/b$.
  - If $a > 0 > b$, then $1/a > 0 > 1/b$.

Applying a Function to Both Sides

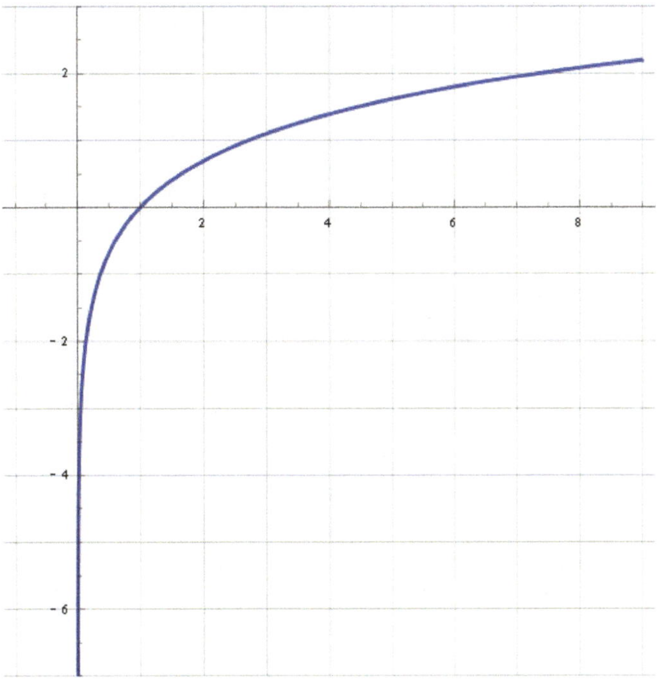

The graph of $y = \ln x$

Any monotonically increasing function may be applied to both sides of an inequality (provided they are in the domain of that function) and it will still hold. Applying a monotonically decreasing function to both sides of an inequality means the opposite inequality now holds. The rules for the additive inverse, and the multiplicative inverse for positive numbers, are both examples of applying a monotonically decreasing function.

If the inequality is strict ($a < b$, $a > b$) *and* the function is strictly monotonic, then the inequality remains strict. If only one of these conditions is strict, then the resultant inequality is non-strict. The rules for additive and multiplicative inverses are both examples of applying a *strictly* monotonically decreasing function.

A few examples of this rule are:

- Exponentiating both sides of an inequality by $n > 0$ when a and b are positive real numbers:

$a \leq b \Leftrightarrow a^n \leq b^n$.

$a \leq b \Leftrightarrow a^{-n} \geq b^{-n}$.

- Taking the natural logarithm to both sides of an inequality when a and b are positive real numbers:

$a \leq b \Leftrightarrow \ln(a) \leq \ln(b)$.

$a < b \Leftrightarrow \ln(a) < \ln(b)$.

This is true because the natural logarithm is a strictly increasing function.

Ordered Fields

If $(F, +, \times)$ is a field and $\leq$ is a total order on F, then $(F, +, \times, \leq)$ is called an ordered field if and only if:

- $a \leq b$ implies $a + c \leq b + c$;
- $0 \leq a$ and $0 \leq b$ implies $0 \leq a \times b$.

Note that both $(\mathbf{Q}, +, \times, \leq)$ and $(\mathbf{R}, +, \times, \leq)$ are ordered fields, but $\leq$ cannot be defined in order to make $(\mathbf{C}, +, \times, \leq)$ an ordered field, because -1 is the square of i and would therefore be positive.

The non-strict inequalities $\leq$ and $\geq$ on real numbers are total orders. The strict inequalities $<$ and $>$ on real numbers are strict total orders.

Chained Notation

The notation $\boldsymbol{a < b < c}$ stands for "$a < b$ and $b < c$", from which, by the transitivity property above, it also follows that $a < c$. Obviously, by the above laws, one can add/subtract the same number to all three terms, or multiply/divide all three terms by same nonzero number and reverse all inequalities according to sign. Hence, for example, $a < b + e < c$ is equivalent to $a - e < b < c - e$.

This notation can be generalized to any number of terms: for instance, $\boldsymbol{a_1 \leq a_2 \leq \ldots \leq a_n}$ means that $a_i \leq a_{i+1}$ for $i = 1, 2, \ldots, n - 1$. By transitivity, this condition is equivalent to $a_i \leq a_j$ for any $1 \leq i \leq j \leq n$.

When solving inequalities using chained notation, it is possible and sometimes necessary to evaluate the terms independently. For instance to solve the inequality $4x < 2x + 1 \leq 3x + 2$, it is not possible to isolate x in any one part of the inequality through addition or subtraction. Instead, the inequalities must be solved independently, yielding $x < 1/2$ and $x \geq -1$ respectively, which can be combined into the final solution $-1 \leq x < 1/2$.

Occasionally, chained notation is used with inequalities in different directions, in which case the meaning is the logical conjunction of the inequalities between adjacent terms. For instance, $a < b = c \leq d$ means that $a < b$, $b = c$, and $c \leq d$. This notation exists in a few programming languages such as Python.

Sharp Inequalities

An inequality is said to be *sharp*, if it cannot be relaxed and still be valid in general. For instance, suppose a is a real number, then the inequality $a^2 \geq 0$ is sharp, whereas the inequality $a^2 \geq -1$ is not sharp.

Inequalities Between Means

There are many inequalities between means. For example, for any positive numbers a_1, a_2, ..., a_n we have $H \leq G \leq A \leq Q$, where

$$H = \frac{n}{1/a_1 + 1/a_2 + \cdots + 1/a_n} \quad \text{(harmonic mean)},$$

$$G = \sqrt[n]{a_1 \cdot a_2 \cdots a_n} \quad \text{(geometric mean)},$$

$$A = \frac{a_1 + a_2 + \cdots + a_n}{n} \quad \text{(arithmetic mean)},$$

$$Q = \sqrt{\frac{a_1^2 + a_2^2 + \cdots + a_n^2}{n}} \quad \text{(quadratic mean)}.$$

Power Inequalities

A "**power inequality**" is an inequality containing a^b terms, where a and b are real positive numbers or variable expressions. They often appear in mathematical olympiads exercises.

Examples

- For any real x,

$$e^x \geq 1 + x.$$

- If $x > 0$ and $p > 0$, then

$$(x^p - 1)/p \geq \ln(x) \geq (1 - 1/x^p)/p.$$

In the limit of $p \to 0$, the upper and lower bounds converge to $\ln(x)$.

- If $x > 0$, then

$$x^x \geq \left(\frac{1}{e}\right)^{1/e}.$$

- If $x \geq 1$, then

$$x^{x^x} \geq x.$$

- If $x, y, z > 0$, then

$$(x+y)^z + (x+z)^y + (y+z)^x > 2.$$

- For any real distinct numbers a and b,

$$\frac{e^b - e^a}{b-a} > e^{(a+b)/2}.$$

- If $x, y > 0$ and $0 < p < 1$, then

$$x^p + y^p > (x+y)^p.$$

- If $x, y, z > 0$, then

$$x^x y^y z^z \geq (xyz)^{(x+y+z)/3}.$$

- If $a, b > 0$, then

$$a^a + b^b \geq a^b + b^a.$$

This inequality was solved by I.Ilani in JSTOR,AMM,Vol.97,No.1,1990.

- If $a, b > 0$, then

$$a^{ea} + b^{eb} \geq a^{eb} + b^{ea}.$$

This inequality was solved by S.Manyama in AJMAA,Vol.7,Issue 2,No.1,2010 and by V.Cirtoaje in JNSA, Vol.4, Issue 2, 130–137, 2011.

- If $a, b, c > 0$, then

$$a^{2a} + b^{2b} + c^{2c} \geq a^{2b} + b^{2c} + c^{2a}.$$

- If $a, b > 0$, then

$$a^b + b^a > 1.$$

This result was generalized by R. Ozols in 2002 who proved that if $a_1, \ldots, a_n > 0$, then

$$a_1^{a_2} + a_2^{a_3} + \cdots + a_n^{a_1} > 1$$

(result is published in Latvian popular-scientific quarterly *The Starry Sky*, see references).

Well-Known Inequalities

Mathematicians often use inequalities to bound quantities for which exact formulas cannot be computed easily. Some inequalities are used so often that they have names:

- Azuma's inequality
- Bernoulli's inequality
- Boole's inequality
- Cauchy–Schwarz inequality
- Chebyshev's inequality
- Chernoff's inequality
- Cramér–Rao inequality
- Hoeffding's inequality
- Hölder's inequality
- Inequality of arithmetic and geometric means
- Jensen's inequality
- Kolmogorov's inequality
- Markov's inequality
- Minkowski inequality
- Nesbitt's inequality
- Pedoe's inequality
- Poincaré inequality
- Samuelson's inequality
- Triangle inequality

Complex Numbers and Inequalities

The set of complex numbers C with its operations of addition and multiplication is a field, but it is impossible to define any relation ≤ so that $(\mathbb{C},+,\times,\leq)$ becomes an ordered field. To make $(\mathbb{C},+,\times,\leq)$ an ordered field, it would have to satisfy the following two properties:

- if $a \leq b$ then $a + c \leq b + c$
- if $0 \leq a$ and $0 \leq b$ then $0 \leq a\, b$

Because ≤ is a total order, for any number a, either $0 \leq a$ or $a \leq 0$ (in which case the first property above implies that $0 \leq -a$). In either case $0 \leq a^2$; this means that $i^2 > 0$ and $1^2 > 0$; so $-1 > 0$ and $1 > 0$, which means $(-1+1) > 0$; contradiction.

However, an operation ≤ can be defined so as to satisfy only the first property (namely, "if $a \leq b$ then $a + c \leq b + c$"). Sometimes the lexicographical order definition is used:

$a \leq b$ if $\mathrm{Re}(a) < \mathrm{Re}(b)$ or $(\mathrm{Re}(a) = \mathrm{Re}(b)$ and $\mathrm{Im}(a) \leq \mathrm{Im}(b))$

It can easily be proven that for this definition $a \leq b$ implies $a + c \leq b + c$.

Vector Inequalities

Inequality relationships similar to those defined above can also be defined for column vectors. If we let the vectors $x, y \in \mathbb{R}^n$ (meaning that $(x_1, x_2, \ldots, x_n)^{\mathrm{T}}$ and $y = (y_1, y_2, \ldots, y_n)^{\mathrm{T}}$ where x_i and y_i we can define the following relationships.

$x = y$ if $x_i = y_i$ for $i = 1, \ldots, n$

$x < y$ if $x_i < y_i$ for $i = 1, \ldots, n$

$x \leq y$ if $x_i \leq y_i$ for $i = 1, \ldots, n$ and $x \neq y$

$x \leqq y$ if $x_i \leq y_i$ for $i = 1, \ldots, n$

Similarly, we can define relationships for $x > y$, $x \geq y$, and $x \geqq y$. We note that this notation is consistent with that used by Matthias Ehrgott in *Multicriteria Optimization* (see References).

The property of trichotomy (as stated above) is not valid for vector relationships. For example, when $x = [2, 5]^{\mathrm{T}}$ and $y = [3, 4]^{\mathrm{T}}$ there exists no valid inequality relationship between these two vectors. Also, a multiplicative inverse would need to be defined on a vector before this property could be considered. However, for the rest of the aforementioned properties, a parallel property for vector inequalities exists.

General Existence Theorems

For a general system of polynomial inequalities, one can find a condition for a solution to exist. Firstly, any system of polynomial inequalities can be reduced to a system of quadratic inequalities by increasing the number of variables and equations (for example by setting a square of a variable equal to a new variable). A single quadratic polynomial inequality in $n - 1$ variables can be written as:

$$X^T AX \geq 0$$

where X is a vector of the variables $X = (x, y, z, \ldots, 1)^T$ and A is a matrix. This has a solution, for example, when there is at least one positive element on the main diagonal of A.

Systems of inequalities can be written in terms of matrices A, B, C, etc. and the conditions for existence of solutions can be written as complicated expressions in terms of these matrices. The solution for two polynomial inequalities in two variables tells us whether two conic section regions overlap or are inside each other. The general solution is not known but such a solution could be theoretically used to solve such unsolved problems as the kissing number problem. However, the conditions would be so complicated as to require a great deal of computing time or clever algorithms.

LIST OF INEQUALITIES

This page lists Wikipedia articles about named mathematical inequalities.

Inequalities in Pure Mathematics

Analysis

- Agmon's inequality
- Askey–Gasper inequality
- Babenko–Beckner inequality
- Bernoulli's inequality
- Bernstein's inequality (mathematical analysis)
- Bessel's inequality
- Bihari–LaSalle inequality
- Bohnenblust-Hille inequality
- Borell–Brascamp–Lieb inequality
- Brezis–Gallouet inequality
- Carleman's inequality
- Chebyshev–Markov–Stieltjes inequalities
- Chebyshev's sum inequality
- Clarkson's inequalities
- Eilenberg's inequality
- Fekete–Szegő inequality
- Fenchel's inequality
- Friedrichs' inequality
- Gagliardo–Nirenberg interpolation inequality
- Gårding's inequality
- Grothendieck inequality
- Grunsky's inequalities
- Hanner's inequalities
- Hardy's inequality
- Hardy–Littlewood inequality
- Harnack's inequality
- Hausdorff–Young inequality
- Hermite–Hadamard inequality
- Hilbert's inequality

- Hölder's inequality
- Jackson's inequality
- Jensen's inequality
- Khabibullin's conjecture on integral inequalities
- Kantorovich inequality
- Karamata's inequality
- Korn's inequality
- Ladyzhenskaya's inequality
- Landau–Kolmogorov inequality
- Lebedev–Milin inequality
- Lieb–Thirring inequality
- Littlewood's 4/3 inequality
- Markov brothers' inequality
- Mashreghi–Ransford inequality
- Max–min inequality
- Minkowski's inequality
- Poincaré inequality
- Popoviciu's inequality
- Prékopa–Leindler inequality
- Rayleigh–Faber–Krahn inequality
- Remez inequality
- Riesz rearrangement inequality
- Schur test
- Shapiro inequality
- Sobolev inequality
- Steffensen's inequality
- Szegő inequality
- Three spheres inequality
- Trace inequalities
- Trudinger's theorem
- Turán's inequalities
- Von Neumann's inequality
- Wirtinger's inequality for functions

Inequalities Relating to Means

- Hardy–Littlewood maximal inequality

- Inequality of arithmetic and geometric means
- Ky Fan inequality
- Levinson's inequality
- Maclaurin's inequality
- Mahler's inequality
- Muirhead's inequality
- Newton's inequalities
- Stein–Strömberg theorem
- Young's inequality

Combinatorics

- Binomial coefficient bounds
- Factorial bounds
- Fishburn–Shepp inequality
- Fisher's inequality
- Ingleton's inequality
- Lubell–Yamamoto–Meshalkin inequality
- Nesbitt's inequality
- Rearrangement inequality
- Schur's inequality
- Stirling's formula (bounds)

Differential Equations

- Grönwall's inequality

Geometry

- Alexandrov–Fenchel inequality
- Aristarchus' inequality
- Barrow's inequality
- Berger–Kazdan comparison theorem
- Blaschke–Santaló inequality
- Bishop–Gromov inequality
- Bogomolov–Miyaoka–Yau inequality
- Bonnesen's inequality
- Brascamp–Lieb inequality

- Brunn–Minkowski inequality
- Castelnuovo–Severi inequality
- Cheng's eigenvalue comparison theorem
- Clifford's theorem on special divisors
- Cohn-Vossen's inequality
- Erdős–Mordell inequality
- Euler's theorem in geometry
- Gromov's inequality for complex projective space
- Gromov's systolic inequality for essential manifolds
- Hadamard's inequality
- Hadwiger–Finsler inequality
- Hinge theorem
- Hitchin–Thorpe inequality
- Isoperimetric inequality
- Jordan's inequality
- Jung's theorem
- Loewner's torus inequality
- Łojasiewicz inequality
- Loomis–Whitney inequality
- Melchior's inequality
- Milman's reverse Brunn–Minkowski inequality
- Minkowski's first inequality for convex bodies
- Myers's theorem
- Noether inequality
- Ono's inequality
- Pedoe's inequality
- Ptolemy's inequality
- Pu's inequality
- Riemannian Penrose inequality
- Toponogov's theorem
- Triangle inequality
- Weitzenböck's inequality
- Wirtinger inequality (2-forms)

Information Theory

- Inequalities in information theory

- Kraft's inequality
- Log sum inequality
- Welch bounds

Algebra

- Abhyankar's inequality
- Pisier–Ringrose inequality

Linear Algebra

- Abel's inequality
- Bregman–Minc inequality
- Cauchy–Schwarz inequality
- Golden–Thompson inequality
- Linear matrix inequality
- Peetre's inequality
- Triangle inequality
- von Neumann's trace inequality
- Weyl's inequality

Number Theory

- Bonse's inequality
- Large sieve inequality
- Pólya–Vinogradov inequality
- Turán–Kubilius inequality
- Weyl's inequality

Probability Theory and Statistics

- Azuma's inequality
- Bennett's inequality, an upper bound on the probability that the sum of independent random variables deviates from its expected value by more than any specified amount
- Bhatia–Davis inequality, an upper bound on the variance of any bounded probability distribution
- Bernstein inequalities (probability theory)
- Boole's inequality

- Borell–TIS inequality
- Burkholder's inequality
- Burkholder–Davis–Gundy inequalities
- Cantelli's inequality
- Chebyshev's inequality
- Chernoff's inequality
- Chung–Erdős inequality
- Concentration inequality
- Cramér–Rao inequality
- Doob's martingale inequality
- Dvoretzky–Kiefer–Wolfowitz inequality
- Eaton's inequality, a bound on the largest absolute value of a linear combination of bounded random variables
- Emery's inequality
- Entropy power inequality
- Etemadi's inequality
- Fannes–Audenaert inequality
- Fano's inequality
- Fefferman's inequality
- Fréchet inequalities
- Gauss's inequality
- Gauss–Markov theorem, the statement that the least-squares estimators in certain linear models are the best linear unbiased estimators
- Gaussian isoperimetric inequality
- Gibbs' inequality
- Hoeffding's inequality
- Hoeffding's lemma
- Jensen's inequality
- Khintchine inequality
- Kolmogorov's inequality
- Kunita–Watanabe inequality
- Le Cam's theorem
- Marcinkiewicz–Zygmund inequality
- Markov's inequality
- McDiarmid's inequality
- Paley–Zygmund inequality
- Pinsker's inequality

- Popoviciu's inequality on variances
- Rao–Blackwell theorem
- Ross's conjecture, a lower bound on the average waiting time in certain queues
- Samuelson's inequality
- Shearer's inequality
- Simpson's paradox
- Talagrand's concentration inequality
- Vitale's random Brunn–Minkowski inequality
- Vysochanskiĭ–Petunin inequality

Topology

- Berger's inequality for Einstein manifolds

Inequalities Particular to Physics

- Ahlswede–Daykin inequality
- Bell's inequality – see Bell's theorem
- Bell's original inequality
- CHSH inequality
- Clausius–Duhem inequality
- Correlation inequality – any of several inequalities
- FKG inequality
- Ginibre inequality
- Griffiths inequality
- Heisenberg's inequality
- Holley inequality
- Leggett–Garg inequality
- Riemannian Penrose inequality
- Rushbrooke inequality
- Sakurai's Bell inequality
- Tsirelson's inequality
- Wigner–d'Espagnat inequality

REFERENCES

1. Hardy, G., Littlewood J.E., Pólya, G. (1999). Inequalities. Cambridge Mathematical Library, Cambridge University Press. ISBN 0-521-05206-8.
2. Beckenbach, E.F., Bellman, R. (1975). An Introduction to Inequalities. Random House Inc. ISBN 0-394-01559-2.
3. Drachman, Byron C., Cloud, Michael J. (1998). Inequalities: With Applications to Engineering. Springer-Verlag. ISBN 0-387-98404-6.
4. Grinshpan, A. Z. (2005), "General inequalities, consequences, and applications", Advances in Applied Mathematics, 34 (1): 71–100, doi:10.1016/j.aam.2004.05.001
5. Murray S. Klamkin. "'Quickie' inequalities" (PDF). Math Strategies.
6. Arthur Lohwater (1982). "Introduction to Inequalities". Online e-book in PDF format.
7. Harold Shapiro (2005,1972–1985). "Mathematical Problem Solving". The Old Problem Seminar. Kungliga Tekniska högskolan. Check date values in: |date= (help)
8. "3rd USAMO". Archived from the original on 2008-02-03.
9. Pachpatte, B.G. (2005). Mathematical Inequalities. North-Holland Mathematical Library. 67 (first ed.). Amsterdam, The Netherlands: Elsevier. ISBN 0-444-51795-2. ISSN 0924-6509. MR 2147066. Zbl 1091.26008.
10. Ehrgott, Matthias (2005). Multicriteria Optimization. Springer-Berlin. ISBN 3-540-21398-8.
11. Steele, J. Michael (2004). The Cauchy-Schwarz Master Class: An Introduction to the Art of Mathematical Inequalities. Cambridge University Press. ISBN 978-0-521-54677-5.

CHAPTER 9

Fundamental Theorems

Chapter Contents

INTRODUCTION

The fundamental theorem of a field of mathematics is the theorem considered central to that field. The naming of such a theorem is not necessarily based on how often it is used or the difficulty of its proofs.

For example, the fundamental theorem of calculus gives the relationship between differential calculus and integral calculus, which are two distinct branches that are not obviously related. Being "fundamental" does not necessarily mean that it is the most basic result. For example, the proof of the fundamental theorem of arithmetic requires Euclid's lemma, which in turn requires Bézout's identity.

The names are mostly traditional, so that for example the fundamental theorem of arithmetic is basic to what would now be called number theory.

The mathematical literature sometimes refers to the fundamental lemma of a field. The term lemma is conventionally used to denote a proven proposition which is used as a stepping stone to a larger result rather than as a useful statement in-and-of itself. The fundamental lemma of a field is often, but not always, the same as the fundamental theorem of that field.

Fundamental Lemmata

- Fundamental lemma of calculus of variations
- Fundamental lemma of Langlands and Shelstad
- Fundamental lemma of sieve theory
- Feinstein's fundamental lemma (information theory)
- Fundamental lemma of interpolation theory (numerical analysis)

Fundamental Theorems of Mathematical Topics

- Fundamental theorem of algebra
- Fundamental theorem of algebraic K-theory
- Fundamental theorem of algebraic number theory
- Fundamental theorem of arithmetic
- Fundamental theorem of calculus
- Fundamental theorem of curves
- Fundamental theorem of cyclic groups
- Fundamental theorem of finitely generated abelian groups
- Fundamental theorem of Galois theory
- Fundamental theorem on homomorphisms
- Fundamental theorem of ideal theory in number fields
- Fundamental theorem of Lebesgue integral calculus
- Fundamental theorem of linear algebra
- Fundamental theorem of linear programming
- Fundamental theorem of projective geometry
- Fundamental theorem of Riemannian geometry
- Fundamental theorem of surfaces
- Fundamental theorem of symmetric polynomials
- Fundamental theorem of topos theory
- Fundamental theorem of ultraproducts
- Fundamental theorem of vector analysis

Non-Mathematical Fundamental Theorems

There are also a number of "fundamental theorems" not directly related to mathematics:

- Fundamental theorem of arbitrage-free pricing
- Fisher's fundamental theorem of natural selection
- Fundamental theorems of welfare economics
- Fundamental equations of thermodynamics
- Fundamental theorem of poker
- Holland's schema theorem, or the "fundamental theorem of genetic algorithms"

LIST OF THEOREMS

This is a **list of theorems**, by Wikipedia page. See also

- Classification of finite simple groups
- List of fundamental theorems
- List of lemmas
- List of conjectures
- List of inequalities
- List of mathematical proofs
- List of misnamed theorems

Most of the results below come from pure mathematics, but some are from theoretical physics, economics, and other applied fields.

A

- AF+BG theorem (*algebraic geometry*)
- ATS theorem (*number theory*)
- Abel's binomial theorem (*combinatorics*)
- Abel's curve theorem (*mathematical analysis*)
- Abel's theorem (*mathematical analysis*)
- Abelian and tauberian theorems (*mathematical analysis*)
- Abel–Jacobi theorem (*algebraic geometry*)
- Abel–Ruffini theorem (*theory of equations*, *Galois theory*)
- Abhyankar–Moh theorem (*algebraic geometry*)
- Abouabdillah's theorem (*geometry,number theory*)
- Absolute convergence theorem (*mathematical series*)
- Acyclic models theorem (*algebraic topology*)
- Addition theorem (*algebraic geometry*)

- Adiabatic theorem (*physics*)
- Ado's theorem (*Lie algebra*)
- Ahiezer's theorem (*complex analysis*)
- Akra–Bazzi theorem (*computer science*)
- Alternate Interior Angles Theorem (*geometry*)
- Albert–Brauer–Hasse–Noether theorem (*algebras*)
- Alchian–Allen theorem (*economics*)
- Alperin–Brauer–Gorenstein theorem (*finite groups*)
- Amitsur–Levitzki theorem (*linear algebra*)
- Analytic Fredholm theorem (*functional analysis*)
- Anderson's theorem (*real analysis*)
- Andreotti–Frankel theorem (*algebraic geometry*)
- Angle bisector theorem (*Euclidean geometry*)
- Ankeny–Artin–Chowla theorem (*number theory*)
- Apéry's theorem (*number theory*)
- Apollonius' theorem (*plane geometry*)
- Appell–Humbert theorem (*complex manifold*)
- Area theorem (conformal mapping) (*complex analysis*)
- Arithmetic Riemann–Roch theorem (*algebraic geometry*)
- Aronszajn–Smith theorem (*functional analysis*)
- Arrival theorem (*queueing theory*)
- Arrow's impossibility theorem (*game theory*)
- Art gallery theorem (*geometry*)
- Artin approximation theorem (*commutative algebra*)
- Artin–Schreier theorem (*real closed fields*)
- Artin–Wedderburn theorem (*abstract algebra*)
- Artin–Zorn theorem (*algebra*)
- Artstein's theorem (*control theory*)
- Arzelà–Ascoli theorem (*functional analysis*)
- Atiyah–Bott fixed-point theorem (*differential topology*)
- Atiyah–Segal completion theorem (*homotopy theory*)
- Atiyah–Singer index theorem (*elliptic differential operators, harmonic analysis*)
- Atkinson's theorem (*operator theory*)
- Aumann's agreement theorem (*statistics*)
- Autonomous convergence theorem (*dynamical systems*)

- Auxiliary polynomial theorem (*diophantine approximation*)
- Ax–Grothendieck theorem (*model theory*)
- Ax–Kochen theorem (*number theory*)
- Aztec diamond theorem (*combinatorics*)

B

- BEST theorem (*graph theory*)
- Babuška–Lax–Milgram theorem (*partial differential equations*)
- Baily–Borel theorem (*algebraic geometry*)
- Baire category theorem (*topology, metric spaces*)
- Balian–Low theorem (*Fourier analysis*)
- Balinski's theorem (*combinatorics*)
- Banach–Alaoglu theorem (*functional analysis*)
- Banach–Mazur theorem (*functional analysis*)
- Banach fixed point theorem (*metric spaces, differential equations*)
- Banach–Steinhaus theorem (*functional analysis*)
- Banach–Stone theorem (*operator theory*)
- Bang's theorem (*geometry*)
- Barbier's theorem (*geometry*)
- Bapat–Beg theorem (*statistics*)
- Baranyai's theorem (*combinatorics*)
- Barwise compactness theorem (*mathematical logic*)
- Bass's theorem (*group theory*)
- Basu's theorem (*statistics*)
- Bauer–Fike theorem (*spectral theory*)
- Bayes' theorem (*probability*)
- Beatty's theorem (*diophantine approximation*)
- Beauville–Laszlo theorem (*vector bundles*)
- Beck's monadicity theorem (*category theory*)
- Beck's theorem (*incidence geometry*)
- Beckman–Quarles theorem (*Euclidean geometry*)
- Beer's theorem (*metric geometry*)
- Behnke–Stein theorem (*several complex variables*)
- Bell's theorem (*quantum theory – physics*)
- Beltrami's theorem (*Riemannian geometry*)
- Belyi's theorem (*algebraic curves*)
- Bendixson–Dulac theorem (*dynamical systems*)

- Berger–Kazdan comparison theorem (*Riemannian geometry*)
- Bernstein's theorem (*functional analysis*)
- Berry–Esséen theorem (*probability theory*)
- Bertini's theorem (*algebraic geometry*)
- Bertrand–Diquet–Puiseux theorem (*differential geometry*)
- Bertrand's ballot theorem (*probability theory, combinatorics*)
- Bertrand's postulate (*prime numbers*)
- Besicovitch covering theorem (*mathematical analysis*)
- Betti's theorem (*physics*)
- Beurling–Lax theorem (*Hardy spaces*)
- Bézout's theorem (*algebraic curves*)
- Bing metrization theorem (*general topology*)
- Bing's recognition theorem (*geometric topology*)
- Binomial inverse theorem (*matrix theory*)
- Binomial theorem (*algebra, combinatorics*)
- Birch's theorem (*Diophantine equation*)
- Birkhoff–Grothendieck theorem (*vector bundles*)
- Birkhoff–Von Neumann theorem (*matrix theory*)
- Birkhoff's representation theorem (*lattice theory*)
- Birkhoff's theorem (*ergodic theory*)
- Birkhoff's theorem (relativity) (*physics*)
- Bishop–Cannings theorem (*economics*)
- Blaschke selection theorem (*geometric topology*)
- Bloch's theorem (*complex analysis*)
- Blondel's theorem (*electric power*) (*physics*)
- Blum's speedup theorem (*computational complexity theory*)
- Bôcher's theorem (*complex analysis*)
- Bogoliubov–Parasyuk theorem (*physics*)
- Bohr–Mollerup theorem (*gamma function*)
- Bohr–van Leeuwen theorem (*physics*)
- Bolyai–Gerwien theorem (*discrete geometry*)
- Bolzano's theorem (*real analysis, calculus*)
- Bolzano–Weierstrass theorem (*real analysis, calculus*)
- Bombieri's theorem (*number theory*)
- Bombieri–Friedlander–Iwaniec theorem (*number theory*)
- Bondareva–Shapley theorem (*economics*)

- Bondy's theorem (*graph theory*, *combinatorics*)
- Bondy–Chvátal theorem (*graph theory*)
- Bonnet theorem (*differential geometry*)
- Boolean prime ideal theorem (*mathematical logic*)
- Borel–Bott–Weil theorem (*representation theory*)
- Borel–Carathéodory theorem (*complex analysis*)
- Borel–Weil theorem (*representation theory*)
- Borel determinacy theorem (*set theory*)
- Borel fixed-point theorem (*algebraic geometry*)
- Borsuk–Ulam theorem (*topology*)
- Bott periodicity theorem (*homotopy theory*)
- Bounded convergence theorem (*measure theory*)
- Bounded inverse theorem (*operator theory*)
- Bourbaki–Witt theorem (*order theory*)
- Brahmagupta theorem (*Euclidean geometry*)
- Branching theorem (*complex manifold*)
- Brauer–Nesbitt theorem (*representation theory of finite groups*)
- Brauer–Siegel theorem (*number theory*)
- Brauer–Suzuki theorem (*finite groups*)
- Brauer–Suzuki–Wall theorem (*group theory*)
- Brauer's theorem (*number theory*)
- Brauer's theorem on induced characters (*representation theory of finite groups*)
- Brauer's three main theorems (*finite groups*)
- Brauer–Cartan–Hua theorem (*ring theory*)
- Bregman–Minc inequality (*discrete mathematics*)
- Brianchon's theorem (*conics*)
- British flag theorem (*Euclidean geometry*)
- Brooks' theorem (*graph theory*)
- Brouwer fixed point theorem (*topology*)
- Browder–Minty theorem (*operator theory*)
- Brown's representability theorem (*homotopy theory*)
- Bruck–Chowla–Ryser theorem (*combinatorics*)
- Brun's theorem (*number theory*)
- Brun–Titchmarsh theorem (*number theory*)
- Brunn–Minkowski theorem (*Riemannian geometry*)
- Buckingham π theorem (*dimensional analysis*)

- Burke's theorem (*probability theory*) (*queueing theory*)
- Burnside theorem (*group theory*)
- Busemann's theorem (*Euclidean geometry*)
- Butterfly theorem (*Euclidean geometry*)

C

- CPCTC (*triangle geometry*)
- Cameron–Martin theorem (*measure theory*)
- Cantor–Bernstein–Schroeder theorem (*Set theory, cardinal numbers*)
- Cantor's intersection theorem (*real analysis*)
- Cantor's theorem (*Set theory, Cantor's diagonal argument*)
- Carathéodory–Jacobi–Lie theorem (*symplectic topology*)
- Carathéodory's existence theorem (*ordinary differential equations*)
- Carathéodory's theorem (*conformal mapping*)
- Carathéodory's theorem (*convex hull*)
- Carathéodory's theorem (*measure theory*)
- Carathéodory's extension theorem (*measure theory*)
- Caristi fixed point theorem (*fixed points*)
- Carleson–Jacobs theorem (*complex analysis*)
- Carlson's theorem (*Complex analysis*)
- Carmichael's theorem (*Fibonacci numbers*)
- Carnot's theorem (*geometry*)
- Carnot's theorem (*thermodynamics*)
- Cartan–Dieudonné theorem (*group theory*)
- Cartan–Hadamard theorem (*Riemannian geometry*)
- Cartan–Kähler theorem (*partial differential equations*)
- Cartan–Kuranishi prolongation theorem (*partial differential equations*)
- Cartan's theorem (*Lie group*)
- Cartan's theorems A and B (*several complex variables*)
- Casey's theorem (*Euclidean geometry*)
- Castelnuovo theorem (*algebraic geometry*)
- Castelnuovo–de Franchis theorem (*algebraic geometry*)
- Castigliano's first and second theorems (*structural analysis*)
- Cauchy integral theorem (*Complex analysis*)
- Cauchy–Hadamard theorem (*Complex analysis*)
- Cauchy–Kowalevski theorem (*partial differential equations*)

- Cauchy's theorem (*geometry*)
- Cauchy's theorem (*finite groups*)
- Cayley–Bacharach theorem (*projective geometry*)
- Cayley–Hamilton theorem (*Linear algebra*)
- Cayley–Salmon theorem (*algebraic surfaces*)
- Cayley's theorem (*group theory*)
- Central limit theorem (*probability*)
- Cesàro's theorem (*real analysis*)
- Ceva's theorem (*geometry*)
- Chasles' theorems
- Chebotarev's density theorem (*number theory*)
- Chen's theorem (*number theory*)
- Cheng's eigenvalue comparison theorem (*Riemannian geometry*)
- Chern–Gauss–Bonnet theorem (*differential geometry*)
- Chevalley's structure theorem (*algebraic geometry*)
- Chevalley–Shephard–Todd theorem (*finite group*)
- Chevalley–Warning theorem (*field theory*)
- Chinese remainder theorem (*number theory*)
- Choi's theorem on completely positive maps (*operator theory*)
- Chomsky–Schützenberger enumeration theorem (*formal language theory*)
- Chomsky–Schützenberger representation theorem (*formal language theory*)
- Choquet–Bishop–de Leeuw theorem (*functional analysis*)
- Chow's theorem (*algebraic geometry*)
- Chowla–Mordell theorem (*number theory*)
- Church-Rosser theorem (*lambda calculus*)
- Clairaut's theorem (*physics*)
- Clapeyron's theorem (*physics*)
- Clark-Ocone theorem (*stochastic processes*)
- Classification of finite simple groups (*group theory*)
- Clausius theorem (*physics*)
- Clifford's circle theorems (*circles*)
- Clifford's theorem on special divisors (*algebraic curves*)
- Closed graph theorem (*functional analysis*)
- Closed range theorem (*functional analysis*)
- Cluster decomposition theorem (*quantum field theory*)
- Coase theorem (*economics*)

- Cochran's theorem (*statistics*)
- Codd's theorem (*relational model*)
- Cohen structure theorem (*commutative algebra*)
- Cohn's irreducibility criterion (*polynomials*)
- Coleman–Mandula theorem (*quantum field theory*)
- Commutation theorem (*von Neumann algebra*)
- Compactness theorem (*mathematical logic*)
- Compression theorem (*computational complexity theory*) (*structural complexity theory*)
- Conley–Zehnder theorem (*dynamical systems*)
- Conservativity theorem (*mathematical logic*)
- Constant rank theorem (*multivariate calculus*)
- Continuous mapping theorem (*probability theory*)
- Convolution theorem (*Fourier transforms*)
- Cook's theorem (*computational complexity theory*)
- Corners theorem (*arithmetic combinatorics*)
- Corona theorem (*Complex analysis*)
- Courcelle's theorem (*graph theory*)
- Cox's theorem (*probability foundations*)
- Craig's theorem (*mathematical logic*)
- Craig's interpolation theorem (*mathematical logic*)
- Cramér's theorem (*statistics*)
- Cramér–Wold theorem (*measure theory*)
- Critical line theorem (*number theory*)
- Crooks fluctuation theorem (*physics*)
- Crossbar theorem (*Euclidean plane geometry*)
- Crystallographic restriction theorem (*group theory, crystallography*)
- Curtis–Hedlund–Lyndon theorem (*cellular automata*)
- Cut-elimination theorem (*proof theory*)
- Cybenko theorem (*neural networks*)

D

- Dandelin's theorem (*solid geometry*)
- Danskin's theorem (*convex analysis*)
- Darboux's theorem (*real analysis*)
- Darboux's theorem (*symplectic topology*)

- Davenport–Schmidt theorem (*number theory, Diophantine approximations*)
- Dawson–Gärtner theorem (*asymptotic analysis*)
- De Branges' theorem (*complex analysis*)
- de Bruijn's theorem (*discrete geometry*)
- De Bruijn–Erdős theorem (incidence geometry)
- De Bruijn–Erdős theorem (graph theory)
- De Finetti's theorem (*probability*)
- De Franchis theorem (*Riemann surfaces*)
- De Gua's theorem (*geometry*)
- De Moivre's theorem (*complex analysis*)
- De Rham's theorem (*differential topology*)
- Deduction theorem (*logic*)
- Denjoy theorem (*dynamical systems*)
- Denjoy–Carleman theorem (*functional analysis*)
- Desargues' theorem (*projective geometry*)
- Descartes' theorem (*plane geometry*)
- Descartes' theorem on total angular defect (*polyhedra*)
- Diller–Dress theorem (*field theory*)
- Dilworth's theorem (*combinatorics, order theory*)
- Dimension theorem for vector spaces (*vector spaces, linear algebra*)
- Dini's theorem (*analysis*)
- Dirac's theorems (*graph theory*)
- Dirichlet's approximation theorem (*Diophantine approximations*)
- Dirichlet's theorem on arithmetic progressions (*number theory*)
- Dirichlet's unit theorem (*algebraic number theory*)
- Disintegration theorem (*measure theory*)
- Divergence theorem (*vector calculus*)
- Dominated convergence theorem (*Lebesgue integration*)
- Donaldson's theorem (*differential topology*)
- Donsker's theorem (*probability theory*)
- Doob decomposition theorem (*stochastic processes*)
- Doob's martingale convergence theorems (*stochastic processes*)
- Doob–Meyer decomposition theorem (*stochastic processes*)
- Dudley's theorem (*probability*)
- Duggan–Schwartz theorem (*voting theory*)
- Dunford–Pettis theorem (*probability theory*)

- Dunford–Schwartz theorem (*functional analysis*)

E

- Earnshaw's theorem (*electrostatics*)
- Easton's theorem (*set theory*)
- Eberlein–Šmulian theorem (*functional analysis*)
- Edge-of-the-wedge theorem (*complex analysis*)
- Edgeworth's limit theorem (*economics*)
- Egorov's theorem (*measure theory*)
- Ehresmann's theorem (*differential topology*)
- Eilenberg–Zilber theorem (*algebraic topology*)
- Elitzur's theorem (*physics*)
- Envelope theorem (*calculus of variations*)
- Equal incircles theorem (*Euclidean geometry*)
- Equidistribution theorem (*ergodic theory*)
- Equipartition theorem (*ergodic theory*)
- Erdős–Anning theorem (*discrete geometry*)
- Erdős–Dushnik–Miller theorem (*set theory*)
- Erdős–Gallai theorem (*graph theory*)
- Erdős–Ginzburg–Ziv theorem (*number theory*)
- Erdős–Kac theorem (*number theory*)
- Erdős–Ko–Rado theorem (*combinatorics*)
- Erdős–Nagy theorem (*discrete geometry*)
- Erdős–Pósa theorem (*graph theory*)
- Erdős–Rado theorem (*set theory*)
- Erdős–Stone theorem (*graph theory*)
- Euclid's theorem (*number theory*)
- Euclid–Euler theorem (*number theory*)
- Euler's polyhedron theorem (*polyhedra*)
- Euler's rotation theorem (*geometry*)
- Euler's theorem (*differential geometry*)
- Euler's theorem (*number theory*)
- Euler's theorem in geometry (*triangle geometry*)
- Euler's theorem on homogeneous functions (*multivariate calculus*)
- Exchange theorem (*linear algebra*)
- Excision theorem (*homology theory*)
- Exterior angle theorem (*triangle geometry*)

- Extreme value theorem (*calculus*)

F

- F. and M. Riesz theorem (*measure theory*)
- FWL theorem (*economics*)
- Faltings' theorem (*diophantine geometry*)
- Farrell–Markushevich theorem (*complex analysis*)
- Fáry's theorem (*graph theory*)
- Fary–Milnor theorem (*knot theory*)
- Fatou's theorem (*complex analysis*)
- Fatou–Lebesgue theorem (*real analysis*)
- Faustman–Ohlin theorem (*economics*)
- Feit–Thompson theorem (*finite groups*)
- Fenchel's duality theorem (*convex analysis*)
- Fenchel's theorem (*differential geometry*)
- Fermat's Last Theorem (*number theory*)
- Fermat's little theorem (*number theory*)
- Fermat's theorem on sums of two squares (*number theory*)
- Fermat's theorem (stationary points) (*real analysis*)
- Fermat polygonal number theorem (*number theory*)
- Fernique's theorem (*measure theory*)
- Ferrero–Washington theorem (*algebraic number theory*)
- Fieller's theorem (*statistics*)
- Final value theorem (*mathematical analysis*)
- Fisher separation theorem (*economics*)
- Fisher–Tippett–Gnedenko theorem (*statistics*)
- Fitting's theorem (*group theory*)
- Five circles theorem (*circles*)
- Five color theorem (*graph theory*)
- Fixed point theorems in infinite-dimensional spaces
- Floquet's theorem (*differential equations*)
- Fluctuation dissipation theorem (*physics*)
- Fluctuation theorem (*statistical mechanics*)
- Ford's theorem (*number theory*)
- Focal subgroup theorem (*abstract algebra*)
- Foster's theorem (*statistics*)
- Four color theorem (*graph theory*)

- Four-vertex theorem (*differential geometry*)
- Fourier inversion theorem (*harmonic analysis*)
- Fourier theorem (*harmonic analysis*)
- Franel–Landau theorem (*number theory*)
- Fraňková–Helly selection theorem (*mathematical analysis*)
- Fredholm's theorem (*Linear algebra*)
- Freidlin–Wentzell theorem (*stochastic processes*)
- Freiman's theorem (*number theory*)
- Freudenthal suspension theorem (*homotopy theory*)
- Freyd's adjoint functor theorem (*category theory*)
- Frobenius determinant theorem (*group theory*)
- Frobenius reciprocity theorem (*group representations*)
- Frobenius theorem (*foliations*)
- Frobenius theorem (*abstract algebras*)
- Froda's theorem (*mathematical analysis*)
- Frucht's theorem (*graph theory*)
- Fubini's theorem (*integration*)
- Fubini's theorem on differentiation (*real analysis*)
- Fuchs's theorem (*differential equations*)
- Fuglede's theorem (*functional analysis*)
- Full employment theorem (*theoretical computer science*)
- Fulton–Hansen connectedness theorem (*algebraic geometry*)
- Fundamental theorem of algebra (*complex analysis*)
- Fundamental theorem of arbitrage-free pricing (*financial mathematics*)
- Fundamental theorem of arithmetic (*number theory*)
- Fundamental theorem of calculus (*calculus*)
- Fundamental theorem on homomorphisms (*abstract algebra*)

G

- Galvin's theorem (*combinatorics*)
- Gauss theorem (*vector calculus*)
- Gauss's Theorema Egregium (*differential geometry*)
- Gauss–Bonnet theorem (*differential geometry*)
- Gauss–Lucas theorem (*complex analysis*)
- Gauss–Markov theorem (*statistics*)
- Gauss–Wantzel theorem (*geometry*)
- Gelfand–Mazur theorem (*Banach algebra*)

- Gelfand–Naimark theorem (*functional analysis*)
- Gelfond–Schneider theorem (*transcendental number theory*)
- Gershgorin circle theorem (*matrix theory*)
- Gibbard–Satterthwaite theorem (*voting methods*)
- Girsanov's theorem (*stochastic processes*)
- Glaisher's theorem (*number theory*)
- Gleason's theorem (*Hilbert space*)
- Glivenko's theorem (*mathematical logic*)
- Glivenko's theorem (*probability*)
- Glivenko–Cantelli theorem (*probability*)
- Goddard–Thorn theorem (*vertex algebras*)
- Gödel's completeness theorem (*mathematical logic*)
- Gödel's incompleteness theorem (*mathematical logic*)
- Godunov's theorem (*numerical analysis*)
- Going-up and going-down theorems (*commutative algebra*)
- Goldberg–Sachs theorem (*physics*)
- Goldie's theorem (*ring theory*)
- Goldstine theorem (*functional analysis*)
- Goldstone theorem (*physics*)
- Golod–Shafarevich theorem (*group theory*)
- Gomory's theorem (*mathematical logic*)
- Goodstein's theorem (*mathematical logic*)
- Gordon–Newell theorem (*queueing theory*)
- Gottesman–Knill theorem (*quantum computation*)
- Gradient theorem (*vector calculus*)
- Graph structure theorem (*graph theory*)
- Grauert–Riemenschneider vanishing theorem (*algebraic geometry*)
- Great orthogonality theorem (*group theory*)
- Green–Tao theorem (*number theory*)
- Green's theorem (*vector calculus*)
- Grinberg's theorem (*graph theory*)
- Gromov's compactness theorem (*Riemannian geometry*)
- Gromov's compactness theorem (*Symplectic topology*)
- Gromov's theorem on groups of polynomial growth (*geometric group theory*)
- Gromov–Ruh theorem (*differential geometry*)
- Gross–Zagier theorem (*number theory*)

- Grothendieck–Hirzebruch–Riemann–Roch theorem (*algebraic geometry*)
- Grothendieck's connectedness theorem (*algebraic geometry*)
- Grötzsch's theorem (*graph theory*)
- Grunsky's theorem (*complex analysis*)
- Grunwald–Wang theorem (*algebraic number theory*)
- Grushko theorem (*group theory*)

H

- H-cobordism theorem (*differential topology*)
- H-theorem (*thermodynamics*)
- Haag's theorem (*quantum field theory*)
- Haag–Łopuszański–Sohnius theorem (*physics*)
- Haboush's theorem (*algebraic groups*, *representation theory*, *invariant theory*)
- Hadamard three-circle theorem (*complex analysis*)
- Hadamard three-lines theorem (*complex analysis*)
- Hadwiger's theorem (*geometry*, *measure theory*)
- Hahn decomposition theorem (*measure theory*)
- Hahn embedding theorem (*ordered groups*)
- Hairy ball theorem (*algebraic topology*)
- Hahn–Banach theorem (*functional analysis*)
- Hahn–Kolmogorov theorem (*measure theory*)
- Hahn–Mazurkiewicz theorem (*continuum theory*)
- Hajnal–Szemerédi theorem (*graph theory*)
- Hales–Jewett theorem (*combinatorics*)
- Hall's marriage theorem (*combinatorics*)
- Halpern–Läuchli theorem (*Ramsey theory*)
- Ham sandwich theorem (*topology*)
- Hammersley–Clifford theorem (*probability*)
- Hardy's theorem (*complex analysis*)
- Hardy–Littlewood maximal theorem (*real analysis*)
- Hardy–Littlewood tauberian theorem (*mathematical analysis*)
- Hardy–Ramanujan theorem (*number theory*)
- Harish–Chandra theorem (*representation theory*)
- Harish–Chandra's regularity theorem (*representation theory*)
- Harnack's curve theorem (*real algebraic geometry*)
- Harnack's theorem (*complex analysis*)
- Hartman–Grobman theorem (*dynamical systems*)

- Hartogs–Rosenthal theorem (*complex analysis*)
- Hartogs' theorem (*complex analysis*)
- Hartogs' extension theorem (*several complex variables*)
- Hasse norm theorem (*number theory*)
- Hasse's theorem on elliptic curves (*number theory*)
- Hasse–Arf theorem (*local class field theory*)
- Hasse–Minkowski theorem (*number theory*)
- Heckscher–Ohlin theorem (*economics*)
- Heine–Borel theorem (*real analysis*)
- Heine–Cantor theorem (*metric geometry*)
- Hellinger–Toeplitz theorem (*functional analysis*)
- Hellmann–Feynman theorem (*physics*)
- Helly–Bray theorem (*probability theory*)
- Helly's selection theorem (*mathematical analysis*)
- Helly's theorem (*convex sets*)
- Helmholtz theorem (classical mechanics) (*physics*)
- Helmholtz's theorems (*physics*)
- Herbrand's theorem (*logic*)
- Herbrand–Ribet theorem (*cyclotomic fields*)
- Higman's embedding theorem (*group theory*)
- Hilbert's basis theorem (*commutative algebra,invariant theory*)
- Hilbert's Nullstellensatz (theorem of zeroes) (*commutative algebra*, *algebraic geometry*)
- Hilbert–Schmidt theorem (*functional analysis*)
- Hilbert–Speiser theorem (*cyclotomic fields*)
- Hilbert–Waring theorem (*number theory*)
- Hilbert's irreducibility theorem (*number theory*)
- Hilbert's syzygy theorem (*commutative algebra*)
- Hilbert's theorem (*differential geometry*)
- Hilbert's theorem 90 (*number theory*)
- Hilbert projection theorem (*convex analysis*)
- Hille–Yosida theorem (*functional analysis*)
- Hindman's theorem (*Ramsey theory*)
- Hinge theorem (*geometry*)
- Hironaka theorem (*algebraic geometry*)
- Hirzebruch signature theorem (*topology*, *algebraic geometry*)

- Hirzebruch–Riemann–Roch theorem (*complex manifolds*)
- Hjelmslev's theorem (*geometry*)
- Hobby–Rice theorem (*mathematical analysis*)
- Hodge index theorem (*algebraic surfaces*)
- Hölder's theorem (*mathematical analysis*)
- Holditch's theorem (*plane geometry*)
- Holland's schema theorem (*genetic algorithm*)
- Holmström's theorem (*economics*)
- Hopf–Rinow theorem (*differential geometry*)
- Hurewicz theorem (*algebraic topology*)
- Hurwitz's automorphisms theorem (*algebraic curves*)
- Hurwitz's theorem (*complex analysis*)
- Hurwitz's theorem (*normed division algebras*)
- Hurwitz's theorem (*number theory*)

I

- Identity theorem (*complex analysis*)
- Identity theorem for Riemann surfaces (*Riemann surfaces*)
- Immerman–Szelepcsényi theorem (*Computational complexity theory*)
- Implicit function theorem (*vector calculus*)
- Increment theorem (*mathematical analysis*)
- Infinite monkey theorem (*probability*)
- Integral root theorem (*algebra, polynomials*)
- Initial value theorem (*integral transform*)
- Integral representation theorem for classical Wiener space (*measure theory*)
- Intermediate value theorem (*calculus*)
- Intercept theorem (*Euclidean geometry*)
- Intersection theorem (*projective geometry*)
- Inverse eigenvalues theorem (*Linear algebra*)
- Inverse function theorem (*vector calculus*)
- Isomorphism extension theorem (*abstract algebra*)
- Isomorphism theorem (*abstract algebra*)
- Isoperimetric theorem (*curves, calculus of variations*)

J

- Jackson's theorem (*queueing theory*)
- Jacobi's four-square theorem (*number theory*)

- Jacobson density theorem (*ring theory*)
- Jacobson–Bourbaki theorem (*algebra*)
- Jacobson–Morozov theorem (*Lie algebra*)
- Japanese theorem for concyclic polygons (*Euclidean geometry*)
- Japanese theorem for concyclic quadrilaterals (*Euclidean geometry*)
- John ellipsoid (*geometry*)
- Jordan curve theorem (*topology*)
- Jordan–Hölder theorem (*group theory*)
- Jordan–Schönflies theorem (*geometric topology*)
- Jordan–Schur theorem (*group theory*)
- Jordan's theorem (multiply transitive groups) (*group theory*)
- Jung's theorem (*geometry*)
- Jurkat–Richert theorem (*analytic number theory*)

K

- Kachurovskii's theorem (*convex analysis*)
- Kanamori–McAloon theorem (*mathematical logic*)
- Kantorovich theorem (*functional analysis*)
- Kaplansky density theorem (*von Neumann algebra*)
- Kaplansky's theorem on quadratic forms (*quadratic forms*)
- Karhunen–Loève theorem (*stochastic processes*)
- Karp–Lipton theorem (*computational complexity theory*)
- Katz–Lang finiteness theorem (*number theory*)
- Kawamata–Viehweg vanishing theorem (*algebraic geometry*)
- Kawasaki's theorem (*paper folding*)
- Kelvin's circulation theorem (*physics*)
- Kempf–Ness theorem (*algebraic geometry*)
- Kharitonov's theorem (*control theory*)
- Khinchin's theorem (*probability*)
- Killing–Hopf theorem (*Riemannian geometry*)
- Kirby–Paris theorem (*proof theory*)
- Kirchhoff's theorem (*graph theory*)
- Kirszbraun theorem (*Lipschitz continuity*)
- Kleene fixed-point theorem (*order theory*)
- Kleene's recursion theorem (*recursion theory*)
- Knaster–Tarski theorem (*order theory*)
- Kneser's theorem (combinatorics)

- Kneser's theorem (differential equations)
- Kochen–Specker theorem (*physics*)
- Kodaira embedding theorem (*algebraic geometry*)
- Kodaira vanishing theorem (*complex manifold*)
- Koebe 1/4 theorem (*complex analysis*)
- Kolmogorov extension theorem (*stochastic processes*)
- Kolmogorov's three-series theorem (*mathematical series*)
- Kolmogorov–Arnold–Moser theorem (*dynamical systems*)
- König's theorem (graph theory) (*bipartite graphs*)
- König's theorem (kinetics) (*physics*)
- König's theorem (*mathematical logic*)
- König's theorem (set theory) (*cardinal numbers*)
- Kövari–Sós–Turán theorem (*graph theory*)
- Kraft–McMillan theorem (*coding theory*)
- Kramers theorem (*physics*)
- Krein–Milman theorem (*mathematical analysis, discrete geometry*)
- Krener's theorem (*control theory*)
- Kronecker's theorem (*diophantine approximation*)
- Kronecker–Weber theorem (*number theory*)
- Krull's principal ideal theorem (*commutative algebra*)
- Krull–Schmidt theorem (*group theory*)
- Kruskal's tree theorem (*order theory*)
- Kruskal–Katona theorem (*combinatorics*)
- Krylov–Bogolyubov theorem (*dynamical systems*)
- Kuhn's theorem (*game theory*)
- Kuiper's theorem (*operator theory, topology*)
- Künneth theorem (*algebraic topology*)
- Kurosh subgroup theorem (*group theory*)
- Kutta–Joukowski theorem (*physics*)
- Kōmura's theorem (*measure theory*)

L

- L-balance theorem (*finite groups*)
- Ladner's theorem (*computational complexity theory*)
- Lafforgue's theorem (*algebraic number theory*)
- Lagrange's theorem (*group theory*)
- Lagrange's theorem (*number theory*)

- Lagrange's four-square theorem (*number theory*)
- Lagrange inversion theorem (*mathematical analysis*, *combinatorics*)
- Lagrange reversion theorem (*mathematical analysis*, *combinatorics*)
- Lambek–Moser theorem (*combinatorics*)
- Lami's theorem (*statics*)
- Landau prime ideal theorem (*number theory*)
- Lasker–Noether theorem (*commutative algebra*)
- Lattice theorem (*abstract algebra*)
- Laurent expansion theorem (*complex analysis*)
- Lauricella's theorem (*functional analysis*)
- Lax–Milgram theorem (*partial differential equations*)
- Lax–Richtmyer theorem (*numerical analysis*)
- Lax–Wendroff theorem (*numerical analysis*)
- Lebesgue covering dimension (*dimension theory*)
- Lebesgue's decomposition theorem (*dimension theory*)
- Lebesgue's density theorem (*dimension theory*)
- Lee Hwa Chung theorem (*symplectic topology*)
- Lebesgue differentiation theorem (*real analysis*)
- Le Cam's theorem (*probability theory*)
- Lee–Yang theorem (*statistical mechanics*)
- Lefschetz fixed-point theorem (*fixed points*, *algebraic topology*)
- Lefschetz–Hopf theorem (*topology*)
- Lefschetz hyperplane theorem (*algebraic topology*)
- Lefschetz theorem on (1,1)-classes (*algebraic geometry*)
- Lehmann–Scheffé theorem (*statistics*)
- Leray's theorem (*algebraic geometry*)
- Leray–Hirsch theorem (*algebraic topology*)
- Lerner symmetry theorem (*economics*)
- Lester's theorem (*Euclidean plane geometry*)
- Levi's theorem (*Lie groups*)
- Levitzky's theorem (*ring theory*)
- Lévy continuity theorem (*probability*)
- Lévy's modulus of continuity theorem (*probability*)
- Lickorish twist theorem (*geometric topology*)
- Lickorish–Wallace theorem (*3-manifolds*)
- Lie's third theorem (*Lie algebra*)

- Lie–Palais theorem (*differential geometry*)
- Lindemann–Weierstrass theorem (*transcendental number theory*)
- Lie–Kolchin theorem (*algebraic groups, representation theory*)
- Liénard's theorem (*dynamical systems*)
- Lindelöf's theorem (*complex analysis*)
- Lindström's theorem (*mathematical logic*)
- Linear congruence theorem (*number theory, modular arithmetic*)
- Linear speedup theorem (*computational complexity theory*)
- Linnik's theorem (*number theory*)
- Lions–Lax–Milgram theorem (*partial differential equations*)
- Liouville's theorem (complex analysis) (*entire functions*)
- Liouville's theorem (conformal mappings) (*conformal mappings*)
- Liouville's theorem (Hamiltonian) (*Hamiltonian mechanics*)
- Löb's theorem (*mathematical logic*)
- Lochs' theorem (*number theory*)
- Looman–Menchoff theorem (*complex analysis*)
- Łoś' theorem (*model theory*)
- Löwenheim–Skolem theorem (*mathematical logic*)
- Lucas' theorem (*number theory*)
- Lukacs's proportion-sum independence theorem (*probability*)
- Lumer–Phillips theorem (*semigroup theory*)
- Luzin's theorem (*real analysis*)
- Lyapunov–Malkin theorem (*stability theory*)
- Lyapunov's central limit theorem (*probability theory*)

M

- M. Riesz extension theorem (*functional analysis*)
- MacMahon Master theorem (*enumerative combinatorics*)
- Mahler's compactness theorem (*geometry of numbers*)
- Mahler's theorem (*p-adic analysis*)
- Maier's theorem (*analytic number theory*)
- Malgrange preparation theorem (*singularity theory*)
- Malgrange–Ehrenpreis theorem (*differential equations*)
- Manin–Drinfeld theorem (*number theory*)
- Mann's theorem (*number theory*)
- Marcinkiewicz theorem (*functional analysis*)
- Marden's theorem (*polynomials*)

- Mazur's control theorem (*number theory*)
- Mergelyan's theorem (*complex analysis*)
- Marginal value theorem (*biology*)
- Markus–Yamabe theorem (*2D stability theory*)
- Martingale representation theorem (*probability theory*)
- Mason–Stothers theorem (*polynomials*)
- Master theorem (*recurrence relations, asymptotic analysis*)
- Maschke's theorem (*group representations*)
- Matiyasevich's theorem (*mathematical logic*)
- Max flow min cut theorem (*graph theory*)
- Max Noether's theorem (*algebraic geometry*)
- Maximal ergodic theorem (*ergodic theory*)
- Maximum power theorem (*electrical circuits*)
- Maxwell's theorem (*probability theory*)
- May's theorem (*game theory*)
- Mazur–Ulam theorem (*normed spaces*)
- Mazur's torsion theorem (*algebraic geometry*)
- Mean value theorem (*calculus*)
- Measurable Riemann mapping theorem (*conformal mapping*)
- Mellin inversion theorem (*complex analysis*)
- Menelaus' theorem (*geometry*)
- Menger's theorem (*graph theory*)
- Mercer's theorem (*functional analysis*)
- Mermin–Wagner theorem (*physics*)
- Mertens' theorems (*number theory*)
- Metrization theorems (*topological spaces*)
- Meusnier's theorem (*differential geometry*)
- Midy's theorem (*number theory*)
- Mihăilescu's theorem (*number theory*)
- Milliken–Taylor theorem (*Ramsey theory*)
- Milliken's tree theorem (*Ramsey theory*)
- Milman–Pettis theorem (*Banach space*)
- Min-max theorem (*functional analysis*)
- Minimax theorem (*game theory*)
- Minkowski's theorem (*geometry of numbers*)
- Minkowski's second theorem (*geometry of numbers*)

- Minkowski–Hlawka theorem (*geometry of numbers*)
- Minlos' theorem (*functional analysis*)
- Miquel's theorem (*circles*)
- Mirsky–Newman theorem (*group theory*)
- Mitchell's embedding theorem (*category theory*)
- Mittag-Leffler's theorem (*complex analysis*)
- Modigliani–Miller theorem (*finance theory*)
- Modularity theorem (*number theory*)
- Mohr–Mascheroni theorem (*geometry*)
- Monge's theorem (*geometry*)
- Monodromy theorem (*complex analysis*)
- Monotone class theorem (*measure theory*)
- Monotone convergence theorem (*mathematical analysis*)
- Montel's theorem (*complex analysis*)
- Moore–Aronszajn theorem (*Hilbert space*)
- Mordell–Weil theorem (*number theory*)
- Moreau's theorem (*convex analysis*)
- Morera's theorem (*complex analysis*)
- Morley's categoricity theorem (*model theory*)
- Morley's trisector theorem (*geometry*)
- Morton's theorem (*game theory*)
- Mostow rigidity theorem (*differential geometry*)
- Mountain pass theorem (*calculus of variations*)
- Moving equilibrium theorem (*economics*)
- Multinomial theorem (*algebra, combinatorics*)
- Multiplication theorem (*special functions*)
- Multiplicity-one theorem (*group representations*)
- Mumford vanishing theorem (*algebraic geometry*)
- Müntz–Szász theorem (*functional analysis*)
- Mycielski's theorem (*graph theory*)
- Myers theorem (*differential geometry*)
- Myhill–Nerode theorem (*formal languages*)

N

- Nachbin's theorem(*complex analysis*)
- Nagata's compactification theorem (*algebraic geometry*)
- Nagata–Smirnov metrization theorem(*general topology*)

- Nagell–Lutz theorem (*elliptic curves*)
- Napoleon's theorem (*triangle geometry*)
- Nash embedding theorem (*differential geometry*)
- Nash–Moser theorem (*mathematical analysis*)
- Newlander–Niremberg theorem (*differential geometry*)
- Newton's theorem about ovals (*curves*)
- Nicomachus's theorem (*number theory*)
- Nielsen fixed-point theorem (*fixed points*)
- Nielsen realization problem (*geometric topology*)
- Nielsen–Schreier theorem (*free groups*)
- Niven's theorem (*mathematics*)
- No cloning theorem (*quantum computation*)
- No free lunch theorem (*philosophy of mathematics*)
- No hair theorem (*physics*)
- No wandering domain theorem (*ergodic theory*)
- No-broadcast theorem (*physics*)
- No-communication theorem (*physics*)
- Noether's theorem (*Lie groups*, *calculus of variations*, *differential invariants*, *physics*)
- Noether's second theorem (*calculus of variations*, *physics*)
- Noether's theorem on rationality for surfaces (*algebraic surfaces*)
- Goddard–Thorn theorem (*vertex algebras*)
- No-trade theorem (*economics*)
- Non-squeezing theorem (*symplectic geometry*)
- Norton's theorem (*electrical networks*)
- Novikov's compact leaf theorem (*foliations*)
- Nyquist–Shannon sampling theorem (*information theory*)

O

- Odd number theorem (*physics*)
- Open mapping theorem (*complex analysis*)
- Open mapping theorem (*functional analysis*)
- Optical equivalence theorem (*quantum optics*) (*quantum physics*)
- Optional stopping theorem (*probability theory*)
- Orbit theorem (*control theory*)
- Orbit-stabilizer theorem (*group theory*)
- Ore's theorem (*graph theory*)

- Orlicz–Pettis theorem (*functional analysis*)
- Ornstein theorem (*ergodic theory*)
- Oseledec theorem (*ergodic theory*)
- Osterwalder–Schrader theorem (*physics*)
- Ostrowski's theorem (*number theory*)
- Ostrowski–Hadamard gap theorem (*complex analysis*)

P

- PCP theorem (*computational complexity theory*)
- Paley's theorem (*algebra*)
- Paley–Wiener theorem (*Fourier transforms*)
- Pandya theorem (*nuclear physics*)
- Pappus's centroid theorem (*geometry*)
- Pappus's hexagon theorem (*geometry*)
- Paris–Harrington theorem (*mathematical logic*)
- Parovicenko's theorem (*topology*)
- Parallel axis theorem (*physics*)
- Parseval's theorem (*Fourier analysis*)
- Parthasarathy's theorem (*game theory*)
- Pascal's theorem (*conics*)
- Pasch's theorem (*order theory*)
- Peano existence theorem (*ordinary differential equations*)
- Peetre theorem (*functional analysis*)
- Peixoto's theorem (*dynamical systems*)
- Penrose–Hawking singularity theorems (*physics*)
- Pentagonal number theorem (*number theory*)
- Perfect graph theorem (*graph theory*)
- Perlis theorem (*graph theory*)
- Perpendicular axis theorem (*physics*)
- Perron–Frobenius theorem (*matrix theory*)
- Peter–Weyl theorem (*representation theory*)
- Phragmen–Lindelöf theorem (*complex analysis*)
- Picard theorem (*complex analysis*)
- Picard–Lindelöf theorem (*ordinary differential equations*)
- Pick's theorem (*geometry*)
- Pickands–Balkema–de Haan theorem (*extreme value theory*)
- Pitman–Koopman–Darmois theorem (*statistics*)

- Pitot theorem (*plane geometry*)
- Pivot theorem (*circles*)
- Planar separator theorem (*graph theory*)
- Plancherel theorem (*Fourier analysis*)
- Plancherel theorem for spherical functions (*representation theory*)
- Poincaré–Bendixson theorem (*dynamical systems*)
- Poincaré–Birkhoff–Witt theorem (*universal enveloping algebras*)
- Poincaré–Hopf theorem (*differential topology*)
- Poincaré duality theorem (*algebraic topology of manifolds*)
- Poincaré recurrence theorem (*dynamical systems*)
- Poisson limit theorem (*probability*)
- Pólya enumeration theorem (*combinatorics*)
- Pompeiu's theorem (*Euclidean geometry*)
- Poncelet's closure theorem (*conics*)
- Poncelet–Steiner theorem (*geometry*)
- Positive energy theorem (*physics*)
- Post's theorem (*mathematical logic*)
- Poynting's theorem (*physics*)
- Preimage theorem (*differential topology*)
- Price's theorem (*physics*)
- Prime number theorem (*number theory*)
- Primitive element theorem (*field theory*)
- Principal axis theorem (*linear algebra*)
- Principal ideal theorem (*algebraic number theory*)
- Prokhorov's theorem (*measure theory*)
- Proper base change theorem (*algebraic geometry*)
- Proth's theorem (*number theory*)
- Pseudorandom generator theorem (*computational complexity theory*)
- Ptolemaios' theorem (*geometry*)
- Pythagorean theorem (*geometry*)

Q

- Quillen–Suslin theorem (*abstract algebra*)
- Quadratic reciprocity theorem
- Quantum threshold theorem (*computer science*) (*theoretical computer science*)

- Quotient of subspace theorem (*functional analysis*)

R

- Rademacher's theorem (*mathematical analysis*)
- Rado's theorem (*harmonic analysis*)
- Radon's theorem (*convex sets*)
- Radon–Nikodym theorem (*measure theory*)
- Raikov's theorem (*probability*)
- Ramanujam vanishing theorem (*algebraic geometry*)
- Ramanujan–Skolem's theorem (*diophantine equations*)
- Ramsey's theorem (*graph theory, combinatorics*)
- Rank-nullity theorem (*linear algebra*)
- Rao–Blackwell theorem (*statistics*)
- Rashevsky–Chow theorem (*control theory*)
- Rational root theorem (*algebra, polynomials*)
- Rationality theorem (*politics*)
- Ratner's theorems (*ergodic theory*)
- Rauch comparison theorem (*Riemannian geometry*)
- Rédei's theorem (*group theory*)
- Reeb sphere theorem (*foliations*)
- Reeh–Schlieder theorem (*local quantum field theory*)
- Reflection theorem (*algebraic number theory*)
- Regev's theorem (*ring theory*)
- Reidemeister–Singer Theorem (*geometric topology*)
- Reider's theorem (*algebraic surfaces*)
- Remmert–Stein theorem (*complex analysis*)
- Residue theorem (*complex analysis*)
- Reversed compound agent theorem (*probability*)
- Reynolds transport theorem (*fluid dynamics*)
- Ribet's theorem (*elliptic curves*)
- Rice's theorem (*recursion theory, computer science*)
- Rice–Shapiro theorem (*computer science*)
- Richardson's theorem (*mathematical logic*)
- Riemann mapping theorem (*complex analysis*)
- Riemann series theorem (*mathematical series*)
- Riemann's existence theorem (*algebraic geometry*)

- Riemann's theorem on removable singularities (*complex analysis*)
- Riemann–Roch theorem (*Riemann surfaces, algebraic curves*)
- Riemann–Roch theorem for smooth manifolds (*differential topology*)
- Riemann–Roch theorem for surfaces (*algebraic surfaces*)
- Riemann singularity theorem (*algebraic geometry*)
- Riesz representation theorem (*functional analysis, Hilbert space*)
- Riesz–Fischer theorem (*real analysis*)
- Riesz–Thorin theorem (*functional analysis*)
- Right-angled Cevian Triangle Theorem (*Advanced Euclidean Geometry*)
- Ringel–Youngs theorem (*graph theory*)
- Robbins theorem (*graph theory*)
- Robertson–Seymour theorem (*graph theory*)
- Robin's theorem (*number theory*)
- Robinson's joint consistency theorem (*mathematical logic*)
- Rokhlin's theorem (*geometric topology*)
- Rolle's theorem (*calculus*)
- Rosser's theorem (*number theory*)
- Rouché's theorem (*complex analysis*)
- Rouché–Capelli theorem (*Linear algebra*)
- Routh's theorem (*triangle geometry*)
- Routh–Hurwitz theorem (*polynomials*)
- Runge's theorem (*complex analysis*)
- Rybczynski theorem (*economics*)
- Ryll–Nardzewski fixed point theorem (*functional analysis*)

S

- S–cobordism theorem (*differential topology*)
- Saccheri–Legendre theorem (*absolute geometry*)
- Sahlqvist correspondence theorem (*modal logic*)
- Saint-Venant's theorem (*physics*)
- Sard's theorem (*differential geometry*)
- Sarkovskii's theorem (*dynamical systems*)
- Savitch's theorem (*computational complexity theory*)
- Sazonov's theorem (*functional analysis*)
- Schaefer's dichotomy theorem (*computational complexity theory*)
- Schauder fixed point theorem (*functional analysis*)
- Schilder's theorem (*stochastic processes*)

- Schnyder's theorem (*graph theory*)
- Schreier refinement theorem (*group theory*)
- Schröder–Bernstein theorems for operator algebras (*operator algebras*)
- Schroeder–Bernstein theorem for measurable spaces (*measure theory*)
- Schur's lemma (*representation theory*)
- Schur's theorem (*Ramsey theory*)
- Schur–Zassenhaus theorem (*group theory*)
- Schwartz kernel theorem (*generalized functions*)
- Schwartz–Zippel theorem (*polynomials*)
- Schwarz–Ahlfors–Pick theorem (*differential geometry*)
- Schwenk's theorem (*graph theory*)
- Scott core theorem (*3-manifolds*)
- Seifert–van Kampen theorem (*algebraic topology*)
- Separating axis theorem (*convex geometry*)
- Shannon–Hartley theorem (*information theory*)
- Shannon's expansion theorem (*Boolean algebra*)
- Shannon's source coding theorem (*information theory*)
- Shell theorem (*physics*)
- Shirshov–Cohn theorem (*Jordan algebras*)
- Shirshov–Witt theorem (*Lie algebras*)
- Shannon's theorem (*information theory*)
- Shift theorem (*differential operators*)
- Siegel–Walfisz theorem (*analytic number theory*)
- Silverman–Toeplitz theorem (*mathematical analysis*)
- Simplicial approximation theorem (*algebraic topology*)
- Sinkhorn's theorem (*matrix theory*)
- Sion's minimax theorem (*game theory*)
- Sipser–Lautemann theorem (*probabilistic complexity theory*) (*structural complexity theory*)
- Siu's semicontinuity theorem (*complex analysis*)
- Six circles theorem (*circles*)
- Six exponentials theorem (*transcendental number theory*)
- Sklar's theorem (*statistics*)
- Skoda–El Mir theorem (*complex geometry*)
- Skolem–Mahler–Lech theorem (*number theory*)
- Skolem–Noether theorem (*simple algebras*)

- Skorokhod's embedding theorem (*statistics*)
- Skorokhod's representation theorem (*statistics*)
- Śleszyński–Pringsheim theorem (*continued fraction*)
- Slutsky's theorem (*probability theory*)
- Smn theorem (*recursion theory, computer science*)
- Sobolev embedding theorem (*mathematical analysis*)
- Sokhatsky–Weierstrass theorem (*complex analysis*)
- Sonnenschein–Mantel–Debreu Theorem (*economics*)
- Sophie Germain's theorem (*number theory*)
- Soul theorem (*Riemannian geometry*)
- Soundness theorem (*mathematical logic*)
- Space hierarchy theorem (*computational complexity theory*)
- Specht's theorem (*matrix theory*)
- Spectral theorem (*functional analysis*)
- Speedup theorem (*computational complexity theory*)
- Sperner's theorem (*combinatorics*)
- Sphere theorem (*Riemannian geometry*)
- Spin-statistics theorem (*physics*)
- Sprague–Grundy theorem (*combinatorial game theory*)
- Squeeze theorem (*mathematical analysis*)
- Stallings theorem about ends of groups (*group theory*)
- Stallings–Zeeman theorem (*algebraic topology*)
- Stanley's reciprocity theorem (*combinatorics*)
- Star of David theorem (*combinatorics*)
- Stark–Heegner theorem (*number theory*)
- Stein–Strömberg theorem (*measure theory*)
- Steiner–Lehmus theorem (*triangle geometry*)
- Steinhaus theorem (*measure theory*)
- Steinitz theorem (*graph theory*)
- Stewart's theorem (*plane geometry*)
- Stinespring factorization theorem (*operator theory*)
- Stirling's theorem (*mathematical analysis*)
- Stokes' theorem (*vector calculus, differential topology*)
- Stolper–Samuelson theorem (*economics*)
- Stolz–Cesàro theorem (*calculus*)
- Stone's representation theorem for Boolean algebras (*mathematical logic*)

- Stone's theorem on one-parameter unitary groups (*functional analysis*)
- Stone–Tukey theorem (*topology*)
- Stone–von Neumann theorem (*functional analysis*, *representation theory* of the *Heisenberg group*, *quantum mechanics*)
- Stone–Weierstrass theorem (*functional analysis*)
- Strassmann's theorem (*field theory*)
- Strong perfect graph theorem (*graph theory*)
- Structure theorem for finitely generated modules over a principal ideal domain (*abstract algebra*)
- Structure theorem for Gaussian measures (*measure theory*)
- Structured program theorem (*computer science*)
- Sturm's theorem (*theory of equations*)
- Sturm–Picone comparison theorem (*differential equations*)
- Subspace theorem (*Diophantine approximation*)
- Supporting hyperplane theorem (*convex geometry*)
- Swan's theorem (*module theory*)
- Sylow theorems (*group theory*)
- Sylvester's determinant theorem (*determinants*)
- Sylvester's theorem (*number theory*)
- Sylvester pentahedral theorem (*invariant theory*)
- Sylvester's law of inertia (*quadratic forms*)
- Sylvester–Gallai theorem (*plane geometry*)
- Symmetric hypergraph theorem (*graph theory*)
- Symphonic theorem (*triangle geometry*)
- Synge's theorem (*Riemannian geometry*)
- Sz.-Nagy's dilation theorem (*operator theory*)
- Szegő limit theorems (*mathematical analysis*)
- Szemerédi's theorem (*combinatorics*)
- Szemerédi–Trotter theorem (*combinatorics*)
- Szpilrajn extension theorem (*axiom of choice*)

T

- Takagi existence theorem (*number theory*)
- Takens' theorem (*dynamical systems*)
- Tameness theorem (*3-manifolds*)
- Tarski's indefinability theorem (*mathematical logic*)
- Taylor's theorem (*calculus*)

- Taylor–Proudman theorem (*physics*)
- Tennenbaum's theorem (*model theory*)
- Thabit ibn Qurra's theorem (*amicable numbers*)
- Thales' theorem (*geometry*)
- The duality theorem (*topology*)
- Thébault's theorem (*geometry*)
- Theorem of de Moivre–Laplace (*probability theory*)
- Theorem of the cube (*algebraic varieties*)
- Theorem of three moments (*physics*)
- Theorem on friends and strangers (*Ramsey theory*)
- Thévenin's theorem (*electrical circuits*)
- Thompson transitivity theorem (*finite groups*)
- Thompson uniqueness theorem (*finite groups*)
- Thue's theorem (*Diophantine equation*)
- Thue–Siegel–Roth theorem (*diophantine approximation*)
- Tietze extension theorem (*general topology*)
- Tijdeman's theorem (*diophantine equations*)
- Tikhonov fixed point theorem (*functional analysis*)
- Time hierarchy theorem (*computational complexity theory*)
- Titchmarsh theorem (*integral transform*)
- Titchmarsh convolution theorem (*complex analysis*)
- Tits alternative (*geometric group theory*)
- Toda's theorem (*computational complexity theory*)
- Tomita's theorem (*operator algebras*)
- Tonelli's theorem (*functional analysis*)
- Topkis's theorem (*economics*)
- Toponogov's theorem (*Riemannian geometry*)
- Torelli theorem (*algebraic geometry*)
- Trichotomy theorem (*finite groups*)
- Trombi–Varadarajan theorem (*Lie group*)
- Trudinger's theorem (*functional analysis*)
- Tsen's theorem (*algebraic geometry*)
- Tunnell's theorem (*number theory*)
- Tutte theorem (*graph theory*)
- Turán's theorem (*graph theory*)
- Turán–Kubilius theorem (*number theory*)

- Tychonoff's theorem (*general topology*)

U

- Ugly duckling theorem (*computer science*)
- Uniformization theorem (*complex analysis, differential geometry*)
- Universal approximation theorem (*neural networks*)
- Universal coefficient theorem (*algebraic topology*)
- Unmixedness theorem (*algebraic geometry*)

V

- Valiant–Vazirani theorem (*computational complexity theory*)
- Van Aubel's theorem (*quadrilaterals*)
- Van der Waerden's theorem (*combinatorics*)
- Van Vleck's theorem (*mathematical analysis*)
- Vantieghems theorem (*number theory*)
- Varignon's theorem (*Euclidean geometry*)
- Vietoris–Begle mapping theorem (*algebraic topology*)
- Vinogradov's theorem (*number theory*)
- Virial theorem (*classical mechanics*)
- Vitali convergence theorem (*measure theory*)
- Vitali covering theorem (*measure theory*)
- Vitali theorem (*measure theory*)
- Vitali–Hahn–Saks theorem (*measure theory*)
- Viviani's theorem (*Euclidean geometry*)
- Von Neumann bicommutant theorem (*functional analysis*)
- Von Neumann's theorem (*operator theory*)
- Von Staudt–Clausen theorem (*number theory*)

W

- Wagner's theorem (*graph theory*)
- Waldhausen's theorem (*geometric topology*)
- Walter theorem (*finite groups*)
- Weber's theorem (*algebraic curves*)
- Wedderburn's little theorem (*ring theory*)
- Wedderburn's theorem (*abstract algebra*)
- Weierstrass–Casorati theorem (*complex analysis*)
- Weierstrass factorization theorem (*complex analysis*)
- Weierstrass preparation theorem (*several complex variables, commutative algebra*)

- Weinberg–Witten theorem (*quantum field theory*)
- Well-ordering theorem (*mathematical logic*)
- Whitehead theorem (*homotopy theory*)
- Whitney embedding theorem (*differential manifolds*)
- Whitney extension theorem (*mathematical analysis*)
- Whitney immersion theorem (*differential topology*)
- Whitney–Graustein Theorem (*algebraic topology*)
- Wick's theorem (*physics*)
- Wiener's tauberian theorem (*real analysis*)
- Wiener–Ikehara theorem (*number theory*)
- Wigner–Eckart theorem (*Clebsch–Gordan coefficients*)
- Wilkie's theorem (*model theory*)
- Wilson's theorem (*number theory*)
- Witt's theorem (*quadratic forms*)
- Wold's theorem (*statistics*)
- Wolstenholme's theorem (*number theory*)

Z

- Z* theorem (*finite groups*)
- ZJ theorem (*finite groups*)
- Zahorski theorem (*real analysis*)
- Zariski's main theorem (*algebraic geometry*)
- Zeckendorf's theorem (*number theory*)
- Zeilberger–Bressoud theorem (*combinatorics*)
- Zsigmondy's theorem (*number theory*)

REFERENCES

- K. D. Joshi (2001). Calculus for Scientists and Engineers. CRC Press. pp. 367–8. ISBN 978-0-8493-1319-6. Retrieved 2009-03-01.

Index